AF430876

THE INDIAN NUMBER SYSTEM
At the Center of the Mathematical World

"The number system, originating from the Indian sub-continent, influenced society like no human artefact. Its basis that lies in a positional number system using symbolic digits, zero-infinity-duality, decimal point, transformed representation of numbers into a scalable, transformable, adaptive form, encoding knowledge like never before, revolutionizing mathematics, logic, computation, natural and social sciences, technology, even poetry, art, philosophy. This offering by my revered teacher, Professor Ganti Prasada Rao, is a scientist-cum-engineer's ode to his motherland's roots – the Indian Number System, from its philosophy, history to modern manifestations. It increases our thirst to fathom its bottomless depth. I enthusiastically recommend it for all, especially students. This book is very interesting and will require more volumes from the author to cover modern computation, hardware, AI / ML, logic, linkages to poetry-philosophy, etc."

- P.P. Chakrabarti,
Professor,
Dept of Computer Science and Engineering
Jointly with Centre of Excellence in Artificial Intelligence and
Ex-Director Indian Institute of Technology Kharagpur.

In this book, Professor Prasada Rao starts with the numbers Zero and One with an easy explanation that even little children can understand and ends up with infinity that puzzles many an elder. He demonstrated very succinctly how the INS led to a range of mathematical concepts which scientists and engineers use as they work on their complex problems. With the clarity with which Rao treats the complex mathematical formulae, the book can shatter the fear of mathematics that make many people in the West become averse to mathematics. Most people deal with numbers but do not realize that the Indian Number System presented a simple formula neatly packaged as a number. This book will help inquisitive children to understand the foundation of mathematics while enabling advanced graduate research students to appreciate the logic of the formulae they use.

- Somayajulu Karamchetty,
Former Senior Manager in the US Army Research Laboratory and
Mentor Emeritus at US SCORE Association.

"Ganti has accomplished a lot in this book. He carefully and methodically develops the origins and concepts behind what is now termed as Hindu-Arabic-Number System or simply referred to as Indian Number System (INS) . What is very significant is the development of connection between the underlying concepts of INS and Linear System Theory concepts of Convolution, Fourier Series, Fourier Transform, Discrete Fourier Transform (and thus Fast Fourier Transform) and Laplace Transform. The fundamental notions of these transforms form the foundation for the modern Communication Theory as well as Control Theory. Indeed, without the quantitative number system, none of the scientific achievements that lead to and still leading the human civilization are possible. Many academic books are not read often, being so specialised and obscure as to appeal only to a handful of possible readers. This book is a treasure to have and read."

- P. Sannuti,
Professor Emeritus,
Department of Electrical & Computer Engineering, Rutgers,
The State University of New Jersey, USA.

"I rarely finish a book in one or two settings. This book on the Indian Number System by Prof. Rao is an exception. By setting things in perspective, it throws new light on a topic about which most of us overly assume that there is nothing more than what is already well known. It is a masterpiece grounded in experience and wisdom, connecting the ancient Indian number system for the first time to some of the fundamental concepts of mathematics developed over the relatively more recent centuries."

- Subbu Allamaraju,
Vice President at Bill.com, USA.

THE INDIAN NUMBER SYSTEM
At the Center of the Mathematical World

Ganti Prasada Rao

Professor Emeritus, Indian Institute of Technology, Kharagpur

BS Publications

An Imprint of **BSP Books Pvt. Ltd.**

4-4-309/316, Giriraj Lane, Sultan Bazar,

Hyderabad - 500 095.

THE INDIAN NUMBER SYSTEM: *At the Center of the Mathematical World*
by Ganti Prasada Rao

Published by:

BSP **BS Publications**

An Imprint of **BSP Books Pvt., Ltd.**

4-4-309/316, Giriraj Lane, Sultan Bazar,

Hyderabad - 500 095

Phone : 040 - 23445688

e-mail : info@bspbooks.net

Website : www.bspbooks.net

ISBN: 978-93-95038-41-6

Contents

CHAPTER 4: THE HINDU DECIMAL NUMBER SYSTEM   27

CHAPTER 5: EVOLUTION OF MODERN FORMS OF REPRESENTATION OF QUANTITIES FROM THE INS   35

CHAPTER 6: RELATIONSHIPS BETWEEN PAIRS OF ARRAYS OF DISCRETE ENTITIES AND OF FUNCTIONS   73

CHAPTER 7: ARRAY-PAIR EVALUATION WITH SHIFTED ARGUMENT   87

CHAPTER 8: REPRESENTATION OF POLYNOMIALS IN MATLAB   97

CHAPTER 9: CONCLUDING REMARKS   99

This work is dedicated in loving memory to my father

Ganti Venkatappadu (1914-1998)

Preface

I have been in the field of education all my life. I went through institutional education as a student and researcher for 22 years starting from 1947, the year of entry into a primary school in the small town Parvatipuram, now in Vizianagaram District of Andhra Pradesh in India. I took my final earned degree of Ph.D. in Electrical Engineering at the Indian Institute of Technology, Kharagpur in 1969. Then, I entered my teaching career first at the PSG College of Technology, Coimbatore, India for two years and moved to IIT Kharagpur in 1971 from where I retired voluntarily in 1997 as I had to take up an assignment as a research advisor in the Engineering Systems Division of the Water and Electricity Department (WED) of the Government of Abu Dhabi in the United Arab Emirates. Having been involved in research on the control of seawater desalination plants in Abu Dhabi for three years, I took up a consultative position in the UNESCO-EOLSS Project in its inception in 1996. Since then I have been actively involved in the development of the Encyclopedia of Life Support Systems (EOLSS: www.eolss.net).

This has been a long period in which I have been travelling extensively on academic missions since 1975 which include Commonwealth Postdoctoral Fellowship to England and many visits to Germany since 1982 as Alexander von Humboldt Research Fellow. I have been attending many international conferences connected with my field of research in System Identification. My wife Meenakshi accompanied me to all the events. In these conferences, I made many friends with whom I still maintain contact. The conferences were all exciting with informal meetings with scholars from various parts of the world at coffee breaks, lunches and dinners. Apparently, due to the good impressions I made in the minds of the participants in these conferences, people used to join us at these events for continued discussions. In addition to answering questions of general nature, my wife and I were happy to share our thoughts on India, and our vegetarian culture and my wife used to enlighten them on the culinary aspects of our food. My friends used to appreciate the colourful silk sarees worn by my wife and informally initiate invitations to us to their countries for future conferences; thus expanding my circle of international friends.

The most important parts of these informal talks in various conferences are queries I used to face on some academic matters from international scholars. These questions were centred on Indians and their contributions to mathematics in ancient times. Many are aware of the comments of admiration from some of the famous mathematicians and scientists in the West.

"It is India that gave us the ingenious method of expressing all numbers using ten symbols each symbol receiving a value of the position and an absolute value; a profound and important idea which appears so simple now that we ignore its

true merit. But its very simplicity and the great ease it lent to our computations put our arithmetic in the first rank of useful inventions; and we shall appreciate the grandeur of the achievement the more when we remember that it escaped the genius of Archimedes and Apollonius, two of the great men produced by antiquity."- Pierre-Simon Laplace

"We owe a lot to the Indians, who taught us how to count, without which no worthwhile scientific discoveries could have been made." Albert Einstein

All of them heard about the Indian contributions of ZERO and NUMBERS and such quotes from the great people of the West. They were curious to know more personally from an Indian whom they assumed to be a mathematician; they were probably misled by my conference presentations and publications that carried a lot of mathematical equations. I am not a mathematician; I simply use some mathematics in my work on electrical engineering, systems and control.

I too heard in my country about zero and the numbers. These seemingly simple questions were quite challenging to me at that time, especially as a scholar facing scholarly audiences. A few people who visited India around that time saw writings on the passing trucks on the roads: '*Mera Bharat Mahaan*' '*My India is great*". I was asked if India was celebrating some great achievements during that time. I was at a loss to give satisfactory answers to such questions.

I used to escape by saying that zero and the decimal number system have spread all over the world making us feel that we cannot live without them. These general statements are not original to me. Many people all over the world say the same. Popular documentaries on the history of numbers by Terry Jones and "What the Ancients knew in India" by Jack Turner assert that the zero and number system from India has "changed the world". These documentaries do not elaborate mathematically any further beyond their role in the simplification of computations and the use of 0 and 1 in modern computers and communications.

What is the significance of my being an Indian if I am unable to say further? In a state of dissatisfaction, I used to return to India to seek help from my mathematician friends in India. All of them were very distinguished scholars in mathematics, but hardly anyone spared enough time to look into the question of zero and the number system beyond what the general public knew. They were busy with their special tracks of study and research. Even those who used to teach the subject of the History of Mathematics were of little help in giving a satisfying justification for the greatness of the zero and number system. On the other hand, I came across people joking that Indian contribution to mathematics is 'ZERO' (implying the literal sense).

Of course, my students at the IIT who join after going through a very highly competitive entrance test conducted at the national level were very brilliant and although they did not ask the same questions, they used to pose some thought-provoking questions, which I consider important. One is about the Laplace transform which in all books is ONLY DEFINED but not derived.

They wished to know how that integral transformation was arrived at. According to the literature, it was suggested by some work of Euler, and Laplace's name was attached to it later. Such queries at home and abroad used to preoccupy me in my free time.

One winter morning, I and my father Ganti Venkatappadu were basking in the sun in the lawn of my quarters B-25 in the IIT Kharagpur campus. I used to read some books outside my profession and my father was reciting some *Vedic* hymns from ancient Indian scriptures to refresh his memory. At one moment I asked him what he was reciting and he said *Upanishads*. He was a Vedic and astrology scholar.

During my childhood, before my admission into primary school, my father taught me orally many things such as arithmetic, 20×20 multiplication tables, and names of the entities of the Hindu calendar and acquainted me with important celestial objects in the night sky, and trained me in useful crafts such as covering books and weaving simple articles with palm reed, and introduced me to many exotic plants of nutritional and medicinal value around our village Kaluvarayi. In the later years after my formal education that culminated in my Ph.D. degree, he guided me and provided some very useful but rare books on some ancient Hindu scriptures to support this effort. My parents lived with us in the IIT Kharagpur campus after my father retired in 1971 as a primary school teacher.

In the early 1980s, I brought a book on nine Upanishads which I bought at Geetha Press, Gorakhpur and began to see what he was reciting out of his memory. It was the *Shanti mantra* from *isha vasyopanishad*. It is a prayer for peace recited in the beginning of that Upanishad. It describes the features of a very special entity- the **Supreme Being**. It does not make any word of request for peace because it is enough to think of the Supreme Being that knows what we want and what we deserve.

OM

PoorNamadah poorNamidam

PoorNaat poorNamudachyatey

PoorNasya poorNamaadaaya

PoorNamevaavasishyatey

Om Santih Santih Santih

Its simplistic meaning is as follows:

(That (The Supreme Being) is FULL, This (Universe) is FULL

FULLNESS arises out of FULLNESS

When the FULLNESS from The FULL is taken away FULLNESS only remains)

This is a prayer addressed to the unique deity the *Supreme Being, Brahma,* and The *Paramatma.*

Also, *Rig-Veda* says **Brahma satyam jagat mithya jeevo brahmaiva na parah**, meaning that *'**the Supreme being is real, all the creation that we see is an illusion, individual self is only Brahma – nothing else'** (Viveka chooDaamaNi).*

Then what about the millions of deities that are often attributed to the Hindu culture? Persons such as Rama and Krishna in India are worshipped by Hindus for their great virtues.

Rama is a role model for a ruler and Krishna teaches the meaning of life and how to lead it. Worship is the expression of the highest degree of respect to all- including our teachers who positively influence our lives. The word 'puujaneeya' meaning 'deserves to be worshipped' is used to express great respect. Recall also the term Hero Worship in our modern social context.

In India, worshipping *vighneSa* the conceptual Lord of Obstacles, annually in *Vinayaka vratam* on a grand scale and also at the beginning of any major endeavour is to express strong commitment to boldly face challenges and impediments on the way towards its completion. Likewise, *Satyanarayana vratam-* Worship of the Lord of Truth, Honesty and Integrity is an oath of commitment to be truthful, and honest. This involves invoking many other divine entities such as the rulers of the eight directions (*ashTa dikpaalakas*), the nine planets (*Nava grahas*), etc. as witnesses. With this clarification, the misconception of polytheism in India may remain in some minds can be cleared.

Some aspects of the discussion on the ancient Indian Knowledge Systems with my father gave me hope that I can proceed to probe further with his help and guidance into the case of ZERO and the Indian Number System. I began to sit with him in his room where he had an unusual collection of books on astrology, *smaartam*, and a set of almanacs since the end of the 1800s. He showed a book which carried short *nighanTus* on Zodiac houses, the 27-star constellations, and numbers. I quickly browsed the small book called *muhurta darpaNam*, in which the initial pages which usually carry publication details were missing. The books were in Sanskrit verses in Telugu script with brief explanations. Going through the *nakshatra nighanTu*, I was able to recall a couple of *Slokas* in the beginning, especially for the star constellations, *aSwani* and *bharaNi*, the "skymarks" (like landmarks on the ground) to identify the zodiac sectors. Of course, only some celestial objects visibly appear in the clear night sky depending on their movements and they are better visible when the moon is not bright. As the season changes other objects take up the visible space in the sky. In my childhood he made me memorize some of these before I was admitted to the school. He used to show the night sky which was free from the glare of the electric lights in our village to locate these constellations. Today in the urban environment, the night sky is not clear due to pollution and the glare of the artificial illumination all around.

After I entered the school, I became busy with my study and school friends. I did not pursue the learning of the ancient Indian texts any further for fear of unemployment and took up the course of modern education which has good job potential. As the eldest son, I realized that I must take up a good job to be able to support my family which includes my younger brother and sisters. I understood the difficulties my father faced in making both ends meet with his meager earnings as a *purOhit* (priest) with his knowledge of the scriptures and occasional work as *purOhit* in weddings and other traditional Hindu rituals and programs. He had to struggle to get a job as a teacher for primary classes which he took in Bobbili and villages around the town. There was no permanence of the job and he was unemployed during the school vacation periods.

In 1953 he finally got the position as a primary school teacher in a village *poorNaSaasanam* which is about 12 miles to the west of Sompeta in Andhra Pradesh. There was no road and therefore no public transport service. In the rainy season, one had to wade through marshes and muddy terrain.

He rode a bicycle with a reinforced fork with two bumpers, a toolbox and a pump. My mother was with us in Sompeta town to facilitate our education. He used to visit us over the weekends. He could carry loads on his bicycle over very rough terrain. His bicycle had two bells to distinguish him from others on the road. He moved all the baggage from *poorNaSaasanam* to Sompeta in batches over the weekends. It is amazing how he managed the distance with no proper track to ride a bicycle. He retired in 1971 and joined my family with my mother, brother and sisters on the IIT campus in Kharagpur in 1976. He was available to help me with information and discussions which supported my investigations on zero and the Indian number system. He passed away on December 20, 1998, at the age of 84.

I began formulating my presentations on the two great Indian contributions to mathematics to be ready to accept invitations for lectures. I was fortunate to receive encouragement at the lectures I gave and my quest to establish the impact of these on the world of mathematics intensified. In 1990 I began giving lectures with increasing material support at many institutions and conferences and I gained enough confidence to face questions from the audiences. On a few occasions when I was hesitant to accept invitations, due to a lack of sufficient new results of research in systems and control to be able to present papers, the organizers insisted on my participation and requested me to repeat what I presented on zero and the INS some years ago at some other events they recollected. I felt immensely encouraged and began to prepare myself with PowerPoint files on my newly acquired laptop. I was able to incorporate some animation features in the mathematical equations enhancing the appeal and clarity. At a few places, I was asked to write a book on the subject of my talks but I could not immediately take up the task due to other regular work. In the years beginning in 2010, organizers of mathematics conferences in India in some places began inviting me to deliver plenary and keynote lectures. One reputed institution in India invited me once in 2018 and again in 2021 to lecture for a Workshop on Ancient Indian Mathematical Traditions.

In the meantime, interest in Ancient Indian Knowledge Systems in India took an important place in the expansion of activities of many Institutions of Eminence in India and Centers devoted to Indian Knowledge Systems was established by the Government of India. The Center of Excellence (CoE) on Indian Knowledge Systems at IIT Kharagpur invited me to make an online presentation on 12th April 2022 and I spoke on "Some Unique Features of Ancient Indian Knowledge Systems". Some of my friends and former students now occupying key positions at IIT Kharagpur suggested that make myself available to help them in the activities of the CoE, especially in the area of mathematics. Having received such a response and encouragement from the academic world over the past three decades, I thought it is time now to put all my thoughts in a comprehensive document which may serve as a textbook.

This work is not on history *per se* of zero and the INS but is an effort to identify the roots of the thinking in the mind of ancient Indians by looking at some samples of ancient literature, education and culture. The precise dates of these inventions and the names of the original inventors will perhaps be never known.

The personal claim for originality as it is widespread in modern times did not exist in ancient India. This urge to claim originality as evidenced by the names of species in the living world in today's biology literature by the names of persons as *extensions to the species names* is conspicuous. In this work, certain dates that are essential for the discussion will be recalled from the literature at appropriate places.

The focus of this work is on the possible influence of the INS concept on the scene of developments in the field of mathematics. An attempt is made to identify the key aspects of similarity, some of which appear superficial and some discernible only at some depth, of many modern mathematical concepts, with the features of the INS. I can make assertions on the impact of the INS based on the timeline of developments in the world of mathematics. First, one observes that there had been no significant developments in the West before the INS was accepted and put to use. Next, all the modern developments that are considered for comparison were long after the INS became an object of wide familiarity. The observations, arguments and comments made in the following are my own, based on, and limited by, my understanding of the wide world of mathematics. I am also unaware of any work that reported on such observations on the INS so far for its impact in detail on the developments in mathematics as has been presented here.

Conventional archaeological methods of digging into the layers of the earth for evidence will not yield any tangible results in cases such as this. This work is an attempt to connect those ancient concepts with some of the later origins now in use in modern mathematics.

I wish to say a few words regarding my choice of words for the title of this book. The origin of ZERO and the Decimal Number System is widely attributed to India. Some in the West refer to the presently used number system as Hindu-Arabic Number System understandably because the system was introduced to the West through the works of the Arabic Scholar Mohammad Ibn Musa Al Khwarizmi. Khwarizmi's book *Kitab al-hisab al-Hindi* on the Calculation with Hindu Numerals was written around 820 CE, translated into Latin as *Algorithmo de Numero Indorum* in the 12th century CE and the symbols were standardized in the 16th century CE. Khwarizmi's *Kitab al-jam wal tariq al-hisab al Hindi* (Addition and subtraction in Indian arithmetic) introduced the decimal positional number system to the Western world. In the Arabic works of Al-Khwarizmi, the reference was only to Indians with the term '*tariq al-hindi*' meaning the *Indian way*, without claiming credit to the Arabs. Also, I find an exception in the way the Arabs write numbers in the system they learnt from the Indians. Arabic text is written from right to left but when a number occurs on the way, the direction is changed- the number is written from left to right. To me, this exception for numbers seems to be full acceptance of the Indian way of handling numbers by the Arabs.

The Latin versions of Al-Khwarizmi's work appear as '*Modus Indorum*' and '*Algorithmo de Numero Indorum*'. Therefore, I dropped the word 'Arabic' from 'Hindu-Arabic' for simplicity in the title of this book. Hindu refers to the country of the river *Sindhu*, whose Western version is the Indus, and 'India' is its derivative. Therefore, the Hindu-Arabic Decimal number system is referred to as the Indian Number System (INS) for the sake of simplicity and it should not give rise to any controversy. Whoever was the true inventor of the system, deserves to be admired for their ingenuity and all of us in the global family should rejoice and take pride in it as a great HUMAN achievement and we should all be grateful to the Arabs for their role in spreading the system.

The mathematical concepts in modern applications are considered in their basic form without too much formal rigour of definitions and descriptions. The terminology is limited to a simple introduction to these concepts to serve our purpose. This is done also to avoid mystifying statements and notation that may distract the readers from the main line of arguments intended to convey the essential message and be comfortable even to readers who use the concepts without requiring detailed formalism. Some interpretations, explanations and illustrative examples are for those who are not specialized in the applications. The presentation considers a wide class of modern concepts for their relation with INS. The presentation of mathematics in this book is deliberately rudimentary since it is addressed to a wide class of readers and many of them may not be specialized in all the aspects covered. The focus is not on mathematics *per se*, but on shades of similarity with the form of the INS. Therefore, the author recommends this work be read without setting too much of a keen eye on hairsplitting distinctions and quibbles for strict formalities in the mathematical definitions and expressions to be able to see a beautiful pattern of ideas that renders the study of mathematics a great pleasure.

This effort had been triggered initially by the queries about India from non-Indian scholars during my visits abroad for research and participation in international conferences and also by some thought-provoking questions from my students in the classroom. It has been regularly supported by the encouragement of my friends and colleagues who were exposed to my ideas at my many lectures in India and abroad. The initial rays of hope for my effort were my father's blessings, discussions, and references. My wife accompanied me to all parts of the world and sat patiently during my talks although the subject was not in her grasp at all she gave useful comments based on her understanding to enhance my presentations. My children Nagalakshmi, Rajeswari and Venkata Lakshmi Narayana (Nani) watched all my online talks and also personally watched me at some conferences and used to provide useful feedback.

Ganti Prasada Rao

Ganti Prasada Rao is an Emeritus Professor of Electrical Engineering, Indian Institute of Technology, Kharagpur, India, with several decades of experience in teaching and research. He is well known for his direct methods of Identification of Continuous-time Systems and applications of special orthogonal functions in the field of Systems and Control. He is an Honorary Professor of East china University of Science and Technology, Shanghai, PR china and was visiting professor at universities in the UK and Europe. He was Technical Advisor to the Water and Electricity Department, Government of Abu Dhabi. He authored/ co-authored over 150 research papers and six books in the field of systems and control. He is on the editorial boards of several International Research Journals. He is a Member of UNESCO-EOLSS Joint Committee. He is a Life Fellow of IEEE and Indian National Academy of Engineering and other professional bodies in India and abroad.

He has been a frequently invited speaker on the Indian mathematical heritage and the impact of the Indian number system on the field of mathematics, subjects on which he has been pursuing independent research over the last 30 years. The results have not been published so far but are put together only in this book for the first time at the behest of many colleagues and friends. This book begins with Aristotle's definition of mathematics as *"the science of quantity"* and reveals many connections between the Indian Number System (INS) and some key concepts in modern mathematics. They are methods of representation and quantitative evaluation of a variety of entities we handle today. We work with them without realizing that conceptually their roots can be traced back to the centuries old structure of the Indian Decimal Number System.

Mathematics, the Science of Quantity

1.1 Introduction

Mathematics has been and will continue to be behind many developments in human capabilities, in the understanding of the real world and the place of humans in it. It helps us to understand what we see around us and to live well with it. It enables us to have some quantitative measure of the entities we come across in the real world, and thereby helps us in making us conscious of our scale and place in the universe, and the way the things happen around us.

To know how many, we are in our families, how old each of the members in them is, how large our dwellings are, how much money we have, how much food and water we have, how hot or cold it is outside, how long we have to wait to get the fruits of a crop, etc. we need quantitative expressions. Many more examples from our daily lives that are connected with quantities and measures can be cited. We get an understanding of the physical world, from the *'values'* of the entities in terms of which we conceive it. The physical world is fitted in a framework of dimensions- space and time, and all these should have a *'measure'* for us to be able to understand. Our needs have 'measures' although our greed may not have any!

According to Aristotle, *"mathematics is the science of quantity"*. Mathematics is the study of measures, relationships, properties of quantities and sets using numbers and symbols. Mathematics is also a universal language with great clarity, definitiveness, brevity and exactitude. Symbols in mathematics are like alphabet, number forms as words having special rules of syntax, expressions and statements bound by logic, and facts stated in theorems which are like poems. Mathematics is a language common to other sciences. *'Number'* is central to mathematics. Its evolution into various forms, its properties, and its role in enabling us to create models – mathematical objects to represent real world situations on our study desks, is what makes it the key element in mathematics.

Like the words in different world languages, numbers had seen a diversity of expressions in the ancient world in different regions. Finally, they found an expression in a system that became unique and universal and the language of mathematics has been standardized. It is the ubiquitous *Hindu-Arabic Number System* or *Indian Number System* (INS). This book is about the number concept in general, and the impact of the INS on the world of mathematics and thereby on other sciences and all aspects of our day-to-day lives in particular. Its focus is on the mathematical 'REPRESENTATION' and the process of 'EVALUATION' of a multitude of entities which are applied in various modern fields of science and technology. Einstein asserted that no worthwhile scientific discoveries could have been made without the INS. In this book, the author presents his own findings which fully justify this sweeping remark from the greatest scientist of all time.

1.1.1   Positive Integers

Small positive integer numbers are recognized even by some animals. A bird may recognize the absence of one egg from its original set of two or three. An animal may have a sense of the number of its offspring. Beyond the small limits, even humans will not be sure unless they keep some record and compare the count with the number of real objects. This trend in record keeping by marks was seen in writing by humans on the Ishango bone. This bone with cut marks was found at the Ishango settlement of fishermen in what is at present the Democratic Republic of Congo in Africa. This is called the 'Ishango bone' and is understood to belong to the Upper Paleolithic era. It is about 10 cm long bone dark brown in color with a piece of quartz attached to it, seemingly for engraving on the bone. The bone has been scratched and polished leaving no clues of the animal to which it belonged. It was probably of a mammal. It is estimated to be 22000 years old. There is also another bone found in Southern Africa which is estimated to be 42000 year old. The cut marks in three sets along its length of the Ishango bone gave rise to different speculations. It could have been a tally stick, or a stick with the marks to facilitate grip or a counting tool. The cuts are in sets of 60s also suggesting some astronomical context in which the stick was used as a counting and recording tool.

Human capability to name numbers developed from humble beginnings. Some communities for example, the Aboriginals in Australia, name numbers as one, two, up to three or four only. Beyond this limit, for them, any number was *many*. The Piraha Amazonian people had no words in consistent use to refer to numbers, even for a single entity. Names for natural numbers were very limited even until many millennia later.

Counting up to ten and twenty for humans became possible probably due to the use of fingers and toes. As names for greater numbers were created, the nomenclature began to follow a systematic approach. The names of numbers literally followed certain pattern of groups such as tens, hundreds, thousands, and so on.

1.1.2   The Symbol for ONE

In the process of recording numbers, a short vertical line, suggesting a cut mark or a single raised finger, became the symbol for ONE in many communities and it remains the same as our modern symbol for ONE. The Babylonians made cuneiform impressions with a wedge like tool on soft clay tablets which were then dried or burned for permanence. They maintained the wedge symbol even for ONE because it lasts better than a vertical line which may be closed or lost in the process of handling soft clay tablets due to its thinness. The same wedge was used for all impressions; only the pressure varied to change the size of the impression. The symbols and the tools used in the ancient communities together with their numeration practices are described an impressive detail by Gullberg (2007) -a medical man with love for mathematics. Terry Jones' popular documentary on 'The Story of 1' (https://vimeo.com/56113926) is an interesting documentary on the history of numbers, in particular of ONE.

1.1.3 Names and Symbols for a Few Initial Natural Numbers

Names for a few initial natural numbers probably came up in a natural way. For example, ONE stands for the moon in the Indian Synonyms due to its uniqueness and regular appearance in the sky. TWO is due to pairs of eyes, arms, ears, etc. In this way, familiarity with objects and organs may have been responsible for the creation of names for TWO. Going further, the fingers of one and both hands were probably behind the creation of numbers FIVE and TEN. Due to the familiarity of these sets of objects, these numbers acquired names. Extensions beyond were mere combinations of the available names. Examples are ONE OVER TEN, TWO OVER TEN, … and so on.

1.1.4 Repeated ONES as Symbols for a FEW Initial Natural Numbers

Having a symbol for ONE suggested repetition of this symbol to denote a few initial numbers. In the Roman system which has been in use until a few centuries ago numbers up to THREE were denoted by sets of repeated ONES. The ancient Egyptian system of numeration had such a direct representation of numbers up to NINE. Such a simple collection of symbols merely becomes a *set* but not a *system*. Larger numbers as repeated ONE symbols become cumbersome to handle occupying increasingly large amount of space for writing. When numbers are written as a sequence of digits in a line, grouping the symbols into the individual digits is a confusing process. Ideally, each of the few initial natural numbers should have individually only one symbol. To have a single symbol for every number indefinitely is another problem especially while writing large numbers. Humans did know this and made some improvements.

Beyond a chosen limit, they began denoting numbers by strings of combinations of basic symbols following certain rules. Some created new symbols for larger numbers. Thus, number systems came into existence.

Table 1.1 Egyptian numerals up to NINE as sets of ONES

ONE	I	SIX	III III
TWO	II	SEVEN	IIII III
THREE	III	EIGHT	IIII IIII
FOUR	IIII	NINE	III III III
FIVE	IIIII		

1.2   Counting in Bunches: The Sense of a Base in Counting

Counting entities one by one is a cumbersome process especially when the total number is large. Suppose we have a bag of grains of rice or wheat. We do not count the number of grains in our taking stock or making trade with them; instead, we measure by volume or weight which combines a number of grains into units of measure used such as liter or kilogram and all our transactions are in these measures and never in the number of grains.

If we have items which are larger than grains, and countable as in the case of medium sized fruits, we may still count them individually. However, in the event of handling large quantities of such items, we resort to making bunches containing a known number of items. They may be bunches of TEN each. Again, these bunches can be counted in larger groups of TEN- each such group contain a TEN of TENS and this process may continue. In this process we get a quicker count of the total number.

Therefore, counting to a familiar limit and then forming a group to be reckoned separately reduces the burden of counting items one-by-one. That familiar number, say TEN in the present case, becomes the basis or base in the method of counting. The orders of the respective groups increase in steps of ONE. In the case groups of TEN, the number of single entities in the first group is TEN, in the next size HUNDRED, THOUSAND in the next and so on. Those left over after the bunching process that are less than TEN and they are termed as UNITS. In a positional system the numbers of the single entities or subgroups from UNITS through TENS, HUNDREDS, .. are written from right to left. A problem arises when some groups vanish in this process of successive bunching. That is, the groups of missing order do not have a way to be shown that they are absent unless we have a symbol denoting ZERO.

Table 1.2 and 1.3 show the ways in which some communities wrote numbers in the ancient times. In most cases, one finds repeated symbols as sets except in Ionian and Hindu systems.

Table 1.2 Twelve written by different communities

People	Expression	Remarks
Babylonians	⟨∇∇	Two symbols for ONE used to denote TWO
Romans	XII	
Egyptians	∩II	
Mayans	•• ⹀	
Greeks (Attic)	ΔII	
Greeks (Ionian)	iβ	Alphabetic numeration system
Hindus	12	Each number has a single unique symbol

Table 1.3 The systems of numeration by different communities

Baby-lonian	Y	YY	YYY	YYYY	YYYYY	YYY YYY	YYYY YYY	YYYY YYYY	YYYY YYYYY	<	
Egyptian	I	II	III	IIII	IIIII	IIIIII	IIIIIII	IIIIIIII	IIIIIIIII	∩	
Greek	A	B	Γ	Δ	E	F	Z	H	θ	I	
Roman	I	II	III	IV	V	VI	VII	VIII	IX	X	
Chinese	一	二	三	四	五	六	七	八	九	十	〇
Mayan	•	••	•••	••••	—	• / —	•• / —	••• / —	•••• / —	— / —	∩ / ∪
Hindu	1	2	3	4	5	6	7	8	9	10	0

Only the INS has ten distinct symbols 0, 1, 2, ….9 without repetitions.

Roman system was in wide use in the west until about 800 years ago. New symbols were created for FIVE (V), TEN (X), FIFTY (L), HUNDRED (C), THOUSAND (M) and so on. The number of symbols increases indefinitely which is a major problem

I II IIIIV V VI VII VIII IX X …… L…C…. M….

Combined symbols are used for numbers around a new symbol with additive and subtractive features in the Roman system: I before the new symbol V denotes subtraction by one, that is, IV denoting FOUR. After the new symbols up to THREE I's are added to denote SIXas VI, SEVEN as VII, and EIGHT as VIII. This rule is followed to denote NINE as IX and so on. This additive and subtractive rule associated with indefinitely increasing burden of new symbols is subjected to conditions which make the Roman numeration quite cumbersome. The Hindu decimal number system appeared to limit the burden to the minimum. The decimal system has been introduced with only ten symbols to represent a number of any size. Hindus also had literal names for many large numbers although they did not use any special symbols for them.

This subtractive rule is applied as in the following

$$IV = 4$$

$$IX = 9$$

$$XL = 40$$

$$XC = 90$$

$$CD = 400$$

$$CM = 900$$

To us this system appears to be quite difficult to understand but the Roman System only was in use till the 13th century CE in the West and an abacus was used as a tool to facilitate calculations. As the numbers grow larger and larger the number of symbols increases and after a limit the Roman alphabet will not be able to provide further symbols; new and weird symbols have to be introduced.

Most number systems of the ancient times carried symbols to denote the sizes of various sets. For all practical purposes, the communities using this system of writing numbers were able to carry on with their use. The Egyptians built great pyramids, the Chinese built the Wall and humans were able to travel great distances and more recently the Romans spread the use of their number system in all walks of life in their domain. The Egyptians used hieroglyph symbols.

1.2.1   The Babylonian System of Numeration

The Babylonian system of numeration (Figure 1.1) is one of the earliest with positional feature. It had *two bases* TEN and SIXTY mixed. Further, there was initially no symbol to identify a blank space until a double wedge symbol came into use. It has no numerical status.

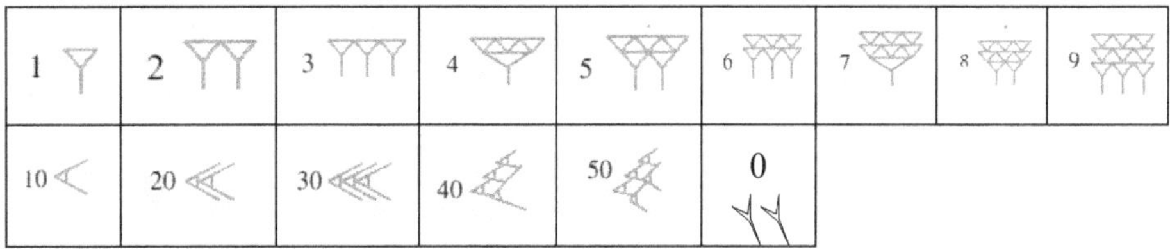

Figure 1.1 The Babylonian system of numeration

This mixture of bases TEN and SIXTY being somewhat difficult for all people to understand and only the scribes could correctly apply this system. The basis of SIXTY is very special in the system and there can be many reasons behind the use of this basis.

1.2.2   The Egyptian Numbers and Symbols

The Egyptians denoted larger numbers pictorially as shown in Figure 1.2.

Stroke	Heelbone	Coiled rope	Lotus flower	Pointing finger	Tadpole	Scribe
1	10	100	1000	10000	100000	1000000

Figure 1.2 Ancient Egyptian number representation

To represent a number less than the basis, there were no special symbols. For instance, FOUR units are denoted by four ONES as **IIII** and FORTY as ∩∩∩∩ and so on. .In the Roman system the same can be noticed with three as the largest before the next symbol appears as **IV** as THREE units are denoted by **III,** and THIRTY is denoted as **XXX.**

1.2.3   Important Features of the INS

Symbols for numbers from ONE to NINE. (1,2,3,4,5,6,7,8.9)

No new symbols for higher numbers; this is the greatest advantage in the writing and understanding numbers. However, large numbers over an extremely large range were given names. However, these names are used in day-to-day activities only up to a limit. But the numbers can be literally read on the basis of the sequence in which the number string is written.

The way we read numbers follows the decimal system and facilitated by the INS.

A symbol for ZERO (0) with the status of a number associated with special rules of arithmetic.

1.2.4   Names of Large Numbers in Sanskrit

The Hindu decimal number system eliminated the burden of additional new symbols to the minimum. It has been introduced with only ten symbols to represent a number of any size. Hindus also had literal names for many large numbers although they did not use any special symbols for them.

Human ability to name numbers progressed. After TEN the names sound as TEN and ONE, TWENTY -ONE, and so on. Indians have been able to name numbers to a great extent. The list of names is given in Table 3. A *crore* which is 10 million is common. Names for further powers of TEN are extensive (Ifrah, 2000). However, names of numbers larger than 10^7 are not used in day-to-life.

Table 1.4 Names of numbers in Sanskrit over a wide range

k	Name of the number of the form 10^k
1	*DaSa*
2	*Sata*
3	*sahasra* (सहस्र)
5	*laksha* (लक्ष)
7	*kOTi* (कोटि)
9	*ayuta* (अयुत)
13	*niyuta* (नियुत)
14	*pakOTi* (पकोटि)
15	*vivAraa* (विवारा)
17	*kshObhyaa* (क्षोभ्या)
19	*vivAhaa* (विवाहा)
21	*kOTippakOTi* (कोटिपकोटी)

Contd...

23	*bahuLa* (बहुल)
25	*naagabaala* (नागाबाला)
28	*naahUTa* (नाहूटा)
29	*teetalambha* (तीतलम्भा)
31	*vyavasthaanaapajnaapati* (व्यवस्थानापज्ञापति)
33	*hetuhilaa* (हेतुहीला)
35	*ninnaahutaa* (निन्नाहुता)
37	*hEtvindriya* (हेत्विन्द्रिय)
39	*samAptalambha* (समाप्तलम्भ)
41	*ganAnAgatee* (गनानागती)
42	*akkhObini* (अक्खोबिनि)
43	*niraavaadya* (निरावाद्य)
45	*mudraabaalaa* (मुद्राबाला)
47	*sarvabaalaa* (सर्वबाला)
49	*bindu* (बिंदु or बिन्दु)
51	*sarvajna* (सर्वज्ञ)
53	*vibhutangamaa* (विभुतनगामा)
56	*abbuda* (अब्बुद)
63	*nirbuddha* (निर्बुद्ध)
70	*ahaahaa* (अहाहा)
77	*abaabaa* (अबाबा).
84	*aTaaTaa* (अटाटा)
91	*sogaangheeka* (सोगान्धीक)
98	*uppala* (उप्पल)
105	*kumuda* (कुमुद)
112	*pundareeka* (पुन्डरीक)
119	*padma* (पद्म)
126	*kathana* (कथन)
133	*mahakathana* (महाकथन)
140	*asankhyEya* (असंख्येय)
421	*dhvajaagraniSamanee* (ध्वजाग्रनिशमनी)

The SAGA of Zero-One of the Greatest Concepts and Scientific Discoveries of all Time

2.1   Introduction

The story of zero is an important episode in the history of mathematics. Much has been written on zero in the literature (Kaplan, 2000; Seife, 2000; Aczel, 2015). The need for such an entity was first felt in many communities just to denote an empty space in the array of symbols in the systems of numeration. When the Sumerians/Babylonians wrote their numbers in cuneiform symbols they used both TEN and SIXTY as bases. Base TEN seems to have been due to the use of fingers but the choice of SIXTY cannot be attributed to such a simple feature of human anatomy as the number of fingers in the human body. It may have been chosen for its special numerical properties- divisibility by many natural numbers or the astronomical significance of the number of days in a solar year. The mixture of the two bases together with the lack of symbol to denote a blank space made it difficult for common people to understand the numbers and the calculations and put the system of numeration at a disadvantage. The blank space between numerals was not so much of a problem as the lack of it after the last numeral which could mean units, tens, hundreds, … and only the context used to help people to see the actual value of the number. To eliminate doubt regarding the presence of a blank space in a position having no numeral, the Babylonians used a double wedge symbol which was just a marker of vacancy and had no role as a numeral in calculations

The Mayans used a symbol which was used in their calendar. However, its use in calculations is not clear and the Mayans being separated by a great ocean from the old world, it had little influence on the developments in mathematics.

Indian texts of 4th century BCE use the Sanskrit word "*Sunya*" to refer to the concept of "void" (zero).

Indian mathematician PingaLa (c. 300 BCE) authored the *chhandaSSastra*, a work on prosody- the patterns of rhyme and rhythm in Sanskrit poetry. In this, he uses zero as a digit (indicated by a dot) and also a description of a binary numeral system. These two digits codify the duration of sounds called *maatra-* as long and short to modulate the pronunciation. Around 50 BC − Indian numerals, evolved from the *braahmi* numerals with the first positional notation and base-ten numeral system began development in India.

Zero which usually means 'nothing' was not considered by many communities in the ancient times. The Greeks left it; because it did not conform to Aristotelian philosophy. They were trapped in the mindset of Geometry. They thought that no geometric shape could represent zero. The Orthodox Church feared it, regarded it as the work of devil and persecuted those who used it.

Zero written in the inscription of a garden of the *caturbhuj* temple in Gwalior was for many years taken as the first ever seen 0 in 876 CE. The rock inscription mentions a garden of 187 *hastas* by 270 *hastas* (1 *hasta* is 1.5ft). Charles Seife (2000) wrote about zero in which a dramatic incident due to the application (division by zero) for a human engineered system was highlighted. Amir Aczel (2015) spent much time and money traveling further east of India in search of zero taking a clue from a French source about the presence of zero in Cambodia and wrote a book on his quest. A zero inscribed on rock was found on one object among the collected ancient monuments destroyed during the violent times in Cambodia and was dated 683 CE, predating the Gwalior inscription by 193 years.

In October 2017 the Bodleian Library in Oxford announced the availability of a manuscript called Bakshali manuscript, a guide to making arithmetical calculations. It was discovered in 1881 in the Bakshali village near Peshawar in modern day Pakistan. Three folios from the manuscript were radio carbon tested and found to belong to three different centuries: from AD 224–383, AD 680–

779, and AD 885–993. This is the oldest seen record of the use of the zero symbol. A large dot, perhaps the precursor of the circle like symbol, is used throughout the manuscript

Archeological research looking for rock inscriptions and ancient writings above and below the layers of the earth can only show a symbol if one is lucky to find an object that carries it. A concept takes many years before entering a record. The Indians carried their knowledge more in their memory than in written documents. Even in the case of business transactions, Indians hardly kept documents and relied mostly on trust as was stated by Megasthenes (c.350BCE-c.290BCE) in his *Indika*. His original is now lost. We now have some excerpts from McCrindle's compiled version (McCrindle, 2008) as quoted by later historians. Therefore, we will try to dig into their thoughts expressed in their prayers, ancient epics and reflected in their socio-cultural practices, to infer their attitude towards ZERO and how they held it in high esteem. The actual origin of the concept should obviously be long before it found such expressions.

2.2   The Indian Reverence to ZERO

The word Hindu we use though out this work does not refer to any branded religion; it refers to the region around and beyond the river Indus. Buddha lived during c.563 BCE-c.480 BCE and his teaching started spreading about the same time. His followers believed that everything has ZERO inherent existence. Even the concept of emptiness is 'devoid of own being'. Buddhism does not have God but Hindu concept of God is as a Supreme Being with unique features as described in the prayers and epics.

In the 2nd century BCE the Roman Empire swallowed up Greece. Fortunately, India was far enough to be insulated from the West. By retaining its freedom of thought, India was able to discover the power of ZERO. The Hindus revered it as divine. For them ZERO evokes images of a primal void. It is linked with OM or AUM- the *Omkaaram*. In the thinking of the Hindus creation and destruction are intermingled. Zero denotes the supreme nothing, lifelessness incarnate, infinite and ineffable, and absolute. Out of the void came the Universe and the Infinite. India actively explored both. It accepted zero and regarded it as a divine incarnation which gave it a great power. Zero and Infinity are self-created and are inseparable. They have no + or − signs. The Supreme Being, the *brahman*, the unique God has no gender. Like the Supreme Being, Zero and Infinity are Complete (*poorNam*). This does not mean being physically full. The Supreme Being is the *largest* and the *smallest*. Its roots stretch back to the dawn of time billions of years before Creation of the World itself. It is nothing but it is everything. These concepts may seem weird to many but not to the Hindu mind.

Think of the absolute zero temperature and absolute zero pressure, two important goals pursued incessantly in the experiments in modern science. Likewise, zero error is another challenge in numerical approximations in computational mathematics. The limitations posed by the word length in our computers remind us of our distance from such a goal. These pursuits will continue forever and there will be no end to them. This is an ultimate challenge posed by this new entity zero.

How the above features are attributed to the Supreme Being (the Brahman) can be seen in the ancient Hindu Scriptures.

Example 2.1: In the *isha vasya upanishad* the *Saanti mantra* (prayer for Peace) is as follows:

OM

poorNamadah poorNamidam

poorNaat poorNamudachyatey

poorNasya poorNamaadaaya

poorNamEvaavaSishyatey

Om Santih Santih Santih

(That (The Supreme Being) is FULL, This (Universe) is FULL

FULLNESS arises out of FULLNESS

When THE FULLNESS from THE FULL is taken away FULLNESS only remains)

If we denote Completeness as f, the above has a mathematical implication $f - f = f$, which is not true for any material thing. It is a unique special property of the Supreme Being. Interestingly, in mathematics it applies only to ZERO and INFINITY (two supernatural entities) and no other worldly quantity. Both are devoid of any sign + or -. They are also inseparable. One arises only from the other.

Example 2.2: *Vishnu sahasranaama stOtram* the Poem in Praise of Lord *vishNu* with His Thousand Names

This finds a place in *anusaanikaparva* of the great ancient epic *mahaabhaarata*. This is a prayer of 149 verses in *anustuup* meter addressing Lord *vishNu* by 1000 chosen names. Sage *vEda vyaasa* authored *mahaabhaarata*.

79

suvarNavarNo hEmangO hvaaraangashchandanaagadee

veeraha vishamah Suunyo ghritaasheer achalaS chalah

75

kaamadevah kaamapaalah kaamee kaantah kritaagamah

anirdesShayavapuuvishnuhveero anantO dhananjayah

Notice the prayer addressing the Lord as *Suunya* (ZERO) and *ananta* (INFINITY) two of the thousand names. The Infinity symbol ∞ according to some people is suggested by the shape of the twisted coil of the great serpent – *anantanaaga*. The Englishman John Wallis (1616 CE-1703 CE) is credited for proposing this symbol. The symbol for ZERO as a circle signifies also endlessness.

Example 2.3: The Prayer by an Elephant King in the Chapter '*gajEndra mOksham*' in the epic *bhaagavatapuraaNa*

An Elephant king with his harem was playing in a pool of water in a forest. A crocodile held him and the rest of the herd ran away.

73

*By whom the **universe is created**;*

Within whom it remains absorbed;

The Supreme Lord,

Who is the Root Cause;

*Who is **devoid of origin, middle and end**;*

Who happens to be everything;

In such a One, the Self-Created; I seek refuge.

Zero (and Infinity) the absolute void out of which all the universe of numbers is created. The universe of numbers is contained between zero and infinity. Zero also absorbs all other numbers (in multiplication); any number multiplied by zero becomes zero itself.

74

I think of the One

That is the universe both exclusive and inclusive of Himself,

The eternal witness to all events, The Spotless and The untarnished,

*The **self-created**.*

The unique entity, remaining *both* within and outside of the universe of other numbers, spotless and untarnished in the sense that you cannot attach anything to IT. Self-created as the original entity. Zero is a supernatural entity unlike any other physical entity which is outside the universe of numbers but is included in it.

83

....The Supreme Lord,

The Complete One: *poorNam*. This term does not mean the physical fullness as we understand. It includes perfection which

The First, The Ineffable,

The Goal of spiritual pursuits,

*The **Complete One**,*

The Supreme Soul,

The Brahma, The Other One,

The physical sense defying,

The Largest and The Smallest - I *propitiate.*

85

*The Lord who is **without gender**,*

***Without any physical form**,*

Not (physically) evidenced by actions properties, differences and places

And yet in the background happens to be all these -I think of Him.

means that its features cannot be surpassed by another entity.

The largest and the smallest: Infinity and ZERO are inseparable. One arises only due to the other and they always appear like an object and its shadow.

Without gender and without any physical form. The mathematical interpretation of this feature-neither male nor female- is that it has no + or − sign.

He fights for release from the crocodile's grip. Despite his own strength and the long fight, he was unable to escape. He became weaker day by day. Then he turns to God- the Supreme Being and prays for his release. That Supreme Being has very unique features that are described in the poems. In fact, the Book itself has a number of stories of devotion to Lord *vishNu/krishNa* but the elephant does not address them but he turns only to the unique Supreme Lord characterized by qualities as detailed in the poem.

Here Prayers addressed to The Supreme in '*Sreemahabhaagavatam*' (Chapter 8, Section 3) by the renowned Telugu poet *Bammera Potana*, reflect these concepts. These are drawn from the very ancient original Sanskrit version *SreemadbhaagavatapuraaNam*. The author attempted to translate those key prayer poems into English. There is the Original Sanskrit version with the same content.

The above features are beyond comprehension for minds that are locked inside the limits of physical entities. By virtue of their thinking that transcends these limits the Hindus were able to integrate their spiritual concepts of the Supreme Being into their world which includes the universe of numbers. They were in a position to assign an appropriate status to this concept of void as a number.

2.3   *Sankhyaa NighanTu*

The ancient Indian Knowledge System was based on memorization and oral transmission. To facilitate memorization the ancient Indian texts were written in verses. Composition of verses in rhyme and meter requires rich vocabulary. For numbers the Indians gave alternative words or synonyms. Synonyms of numbers are provided in the verses in the *sankhyaa nighanTu*. This is a small book describing synonyms for use in place of numbers. These are chosen from the real world objects occurring in sets of one, two, three,… nine. There are also synonyms for ZERO. Table 2.1 lists verses describing the synonyms for various numbers in *sankhyaa nighanTu*.

Indian Mathematician Brahmagupta (c.598 CE-c.668 CE) gave ZERO the right status as a number. In his book *Brahma-sphuta-siddhanta*, he clearly explained the use of zero not only as a placeholder but also as a digit, and explained the INS. Indian mathematician Aryabhata (476 CE-550 CE) was the first to refer to the digit '0' as '*kha*' meaning the sky. Bhaskara II (1114 CE-1185 CE) uses the term *khahaara-* meaning division by zero and its consequences producing larger and larger results; only zero can generate infinity and vice versa by division. He was the first to propose infinitesimals laying foundations for differential calculus centuries before Newton. His work is the famous *siddhantha SiromaNi*.

Table 2.1 *sankhyaa nighanTu*

Number	Verse providing the synonyms
1	*SaSee sOmaSSaSaankaScha induSchandrah kaLaanidhih raajaa vidhu ssudhanSuS cha yama EkO javastadhaah* All these words refer to the moon which is the ONE object nearest to us and conspicuously visible to us.
2	*akshi chakshuh karO nEtram lOchanam baahu karNakaah paksha drishTi dwayam yugma mambakou nayanekshaNe* These terms refer to organ *pairs* such as eyes, arms, hands,... and the numerical words *dwayam* and *yugmam* mean pair.
3	*vahnee raamaSSikhee chaagnih paavakO dahana analou Sankarraakshi puree lOkaastreeni kaala strayO guNah* These words are used to refer to three- fire, combustion, the fire emitting third eye of the three eyed *Siva*, and the three *gunas- satva, rajas* and *tamas* the three features of nature
4	*abdhi saagara chatvaari vanaraaSir yugOmbudhih chaturvaardhir gatischaapi jaladhir neeradhistadha* *chatvaari, chatur* are the numeral names and others which have the same number of objects – the oceans (four- east, west, north, south), the four *yugas*.
5	*indryam panchamam jnana mishu baaNascha maargaNah vratou bhuutam Sharah parvaa praaNasya vishayastatha* Those entities forming sets of five- senses, the five arrows of cupid, five elements of nature, etc. *pancha*- the name of the number
6	*Saastram shaTcha ruchiSchaiva kaalaScha ritusangnikam rasadravyamcha koSaScha shaD darsana shaDaagamaa* Sets of six- *Saastras*, culinary tastes, times, seasons, etc. *shaT*- the name of the number
7	*Sailodrir dweepa vaayuScha muni saptaachalo girih turagaaSwa nagaa gothraa maheedhara rishi sangnakah* Sets of seven, hills, islands, gases, sages, the horses of the sun etc. *sapta* is the name of the number
8	*ashTamam gaja karNeecha diggajou danthi hasthinou saamajo mattamaatangah dikpaala vasu vaaranaah* *ashTamam*- the number, Sets of eight- the elephants bearing the eight directions, the rulers of the eight directions, *vasus*, etc.
9	*navamam navaratnamcha brahmaacha kamalaasanah nidhir grahaScha khanDancha randhro bhaavaScha labdhakah* *navamam*- number nine. the nine jewels, nine treasures, planets, nine holes in the body, etc.
0	*aakaaSam gaganam Suunyamantariksham marutpatham* The word *Suunyam* is the most common in our regular use. The other words refer to the sky and outer space entities that evoke a sense of vastness and indicate the Hindu reverence for ZERO.

Some of the alphabets are also used to denote numbers:

kaadirnavo Taadir navah paadi panchacha yaashTakam

1-9 from *ka*, 1-9 from *Ta*, 1-5 from *pa*, 1-8 from *ya*.

(Source: *'sankhyaa nighanTu'* in *'muhuurta darpaNam'*.

nighanTu: A book of synonyms or Thesaurus (Dictionary not in alphabetical organization of words but as sets of synonyms for each word in the form of poems), ***muhuurta darpaNam***: Mirror of Moments (of time))

The Evolution of the Number Concept

3.1 Introduction

Positive integers are the first numbers conceived by humans since they denote discrete visible entities. The numbers one, two, three, etc. are called natural numbers and the entire set of natural numbers is denoted by $\mathbb{N}$. Fractions represent the parts of a whole or collection of objects. Subsequently such portions became necessary and the full entities are divided into parts and later regarded as practically relevant numbers. Positive integers and fractions together are denoted as 'positive real numbers'.

3.2 Negative Numbers and the Real Line

The concept of a 'negative number' was beyond the ability of many communities but the Indians visualized a negative number as one that denotes a 'borrowed' entity. In Sanskrit, '*rin*' which means debt, became an attribute to a negative number which mathematically is now denoted by a '—' sign. As a consequence of this new form of numbers, the positive quantities are referred to by the term 'wealth' or '*dhan*' in Sanskrit and + sign is used as the counterpart of the minus sign. In actual practice the + sign is not added in front of positive numbers when they stand alone but in the presence of negative numbers, the +sign is used to distinguish them from the negative quantities. Thus the 'Real Line', a horizontal line extending from far left to the far right became a representation for real numbers. There is one point on the real line that separates the positive and negative sides which is neither − nor + and it is ZERO. Zero had an interesting history before it became a member of the number family. Hindus are credited for boldly giving it the status of a number. While some communities avoided zero since it means nothing and there was no need to consider nothing.

3.3   Imaginary Numbers

Returning to the evolution of numbers after the formation of the real line, calculations using real numbers followed for many years in the ancient times. The square root of a positive number is recognized to exist- in fact two square roots exist, one positive and the other its mirror reflection on the left side of zero on the real line. Square roots are visualized for all positive numbers up to zero. When attempts were made to determine the square roots of negative numbers, mathematicians could not visualize anything on the real line which is the only number space established until the $j = \sqrt{-1}$ was introduced. Then the square roots of negative numbers are calculated following the procedure for positive numbers and the symbol $\pm j$ is attached to the results. As it was not possible to conceive these results on the real line, these results are called 'imaginary'. Thus imaginary numbers were created and the symbol $\pm j$ was give the name 'operator' which rotates the position on the real line by $\pm\dfrac{\pi}{2}$ or $\pm 90°$. While multiplication of two real numbers gives results on the real line itself, multiplication of a real number with one associated with $+j$ gives the same value which does not exist on the real line but on a line perpendicular to the real line at zero towards the north. Likewise, multiplication with a number associated with $-j$ give the result on the vertical line at zero towards the south. The question of existence of roots is now resolved. Square roots exist for negative numbers too but you will find them on the vertical line at zero which is called the 'imaginary axis'. A number with real and imaginary parts is called 'complex' and consequently a number in general attain *two dimensions*. Figure 3.1 shows the evolution of a number into a two-dimensional entity.

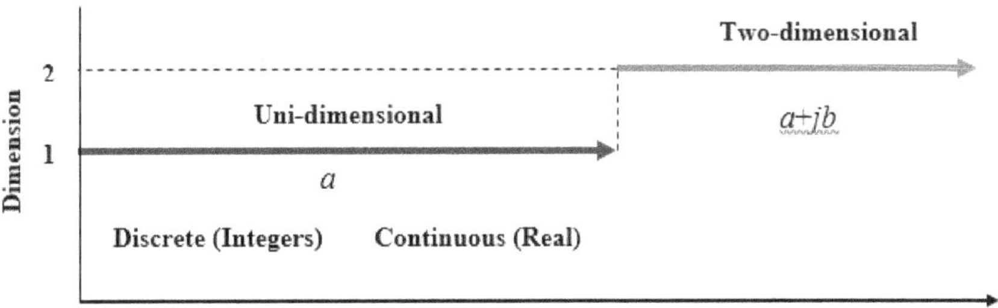

Figure 3.1 Development of the number concept from one dimension to two

3.4   The General Number Space

In the number plane of two dimensions with which we usually work we mark four points $+\infty, -\infty, +j\infty, -j\infty$. In fact there is only one infinity ∞ diametrically opposite to the point ZERO in this number space which is like Riemann space. The associated symbols specify the direction in which one may proceed towards this unique entity which has certain features akin to those of ZERO. The general number space is shown in Figure 3.2.

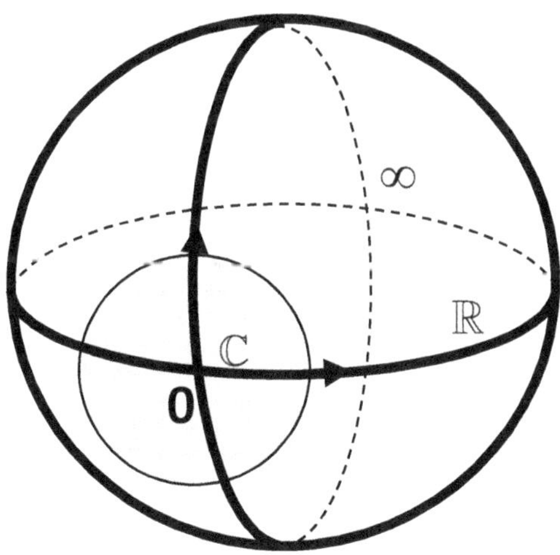

Figure 3.2 The Number Space as Riemann Surface of infinite sphere

The Hindu Decimal Number System

4.1   Timeline of Developments

The timeline of developments of the **Hindu Decimal Number System** or Hindu-Arabic Numeration System or Indian Number System (INS) is given in Table 4.1.

Table 4.1 Chronology of significant developments in mathematics in the current era culminating in the appearance of the INS

CE	Event
300	The earliest known use of zero as a decimal digit is introduced by Indian mathematicians.
c. 400	Bakhshali manuscript is written by Jaina mathematicians, describes a theory of the infinite containing different levels of infinity
550	Hindu mathematicians give zero a numeral representation in the positional notation Indian numeral system.
628	Brahmagupta's *Brahma-sphuta-siddhanta*, explains zero clearly and develops the INS . It also gives rules for handling both negative and positive numbers,
773	Brahmagupta's *Brahma-sphuta-siddhanta* was brought to Baghdad by Kanka, a scholar from Ujjain, to explain the Indian system of arithmetic astronomy and the Indian numeral system. Al-Fazari translated the *Brahma-sphuta-siddhanta* into Arabic upon the request of the Abbasid Khalifa, Abu Jafar Al Mansoor.
820	Mahammad Inbn Musa Al-Khwarizmi – a Persian mathematician, authored *Kitab al jam wa-l tariq al-hisab al Hind* the Al-Jabr and translations of his book on arithmetic introduce the Hindu–Arabic

Contd…

CE	Event
	decimal number system to the Western world in the 12th century. The term algorithm is also named after him. Advances enabled by the decimal system led to its standard use throughout the region and the world.
c. 1000	Pope Sylvester II introduces the abacus using the Hindu–Arabic numeral system to Europe.
1030	Ali Ahmad Nasawi authored a treatise on the decimal and sexagesimal number systems.
1202	Leonardo Fibonacci demonstrates the utility of Hindu–Arabic numerals in his Liber Abaci (Book of the Abacus)

The INS first came from India to the House of Wisdom in Baghdad, working under the patronage of Caliph Al Mansour. A Persian scholar Md. Ibn-Musa Al Khwarizmi's wrote a famous manual on mathematics in which he introduced the INS. It was translated into Latin in the 12th Century and the symbols were standardized in the 16th Century.

At a time when the West was using the Roman number system, the INS came to North Africa with the Arab Traders. It was essentially used in trade and commerce but it was not welcome by the government of Italy which banned the use of the new system. The traders however secretly used the INS for its efficiency and simplicity and maintained accounts separately but prepared them in parallel in the Roman system for submission to official agencies for audit and taxation.

An Italian mathematician, Leonardo Bonacci, Leonardo of Pisa or Leonardo Bigollo Pisano popularly known as Fibonacci, who lived in North Africa looked at the INS in use in the markets in Algeria was greatly impressed by the number system and the ease with which calculations were made accurately. He wrote a book -*Liber Abaci* which was published in 1202. He described the INS and its use in a chapter- *Modus Indorum* in this book. This triggered the interest in the INS in the West.

4.2   The General Format of Number System-
The Modern System and its Features

An integral number is written as a horizontal array of numerals from the set [1, 2, ..., 9,0]. The digit on the extreme right denotes ONES; that is, the weight of the digit is ONE. As we move to the left the weight increases ten times from the previous position. The weights from right to left are 1, 10, 100, 1000, ... and so on. Aryabhata (5th century AD) puts this in Sanskrit as: '*sthanaat sthaanam daSaguNam syat*' (from position to position the value is ten times). If we represent a number as an array of digits belonging to the set [1, 2, ..., 9,0], there is a corresponding weighting sequence or an array of weights associated with this array but it is not written but understood. For numbers which include fractions a dot separates the integer part from the decimal fraction written on the right side. The organization of the INS is shown in Table 4.2

Table 4.2 The general format of the INS

	Integer part			Separator decimal point	Fractional part		
Position	---	2	1	0		-1	-2	---
The number written as	---	d_2	d_1	d_0	$\bullet$	d_{-1}	d_{-2}	---
Weight for each position	---	100 10^2	10 10^1	1 10^0		1/10 10^{-1}	1/100 10^{-2}	---

In general the weight for the kth position is 10^k. A number is written in the INS form as

$$\cdots d_2 d_1 d_0 \bullet d_{-1} d_{-2} \cdots$$

or

$$\cdots d_k \cdots, \quad k = -\infty \rightarrow +\infty \text{ is simply an array of from } [1,2,3,\ldots 9,0].$$

The number of digit is equal to TEN since this is a decimal system of system of base TEN. The base is also called the RADIX. The array of weights associated with the above sequence is implied and is $\cdots 10^2 10^1 10^0 \bullet 10^{-1} 10^{-2}$.

The VALUE of the number is then obtained by the *weighted sum of the digits*:

$$N = \cdots \left(10^2 \times d_2\right) + \left(10^1 \times d_1\right) + \left(10^0 \times d_0\right) + \bullet \left(10^{-1} \times d_{-1}\right) + \left(10^{-2} \times d_{-2}\right) \cdots$$

or

$$N = \sum_{k=-\infty}^{\infty} 10^k d_k \ .$$

This can be generalized for a system of base or radix r as

$$N = \sum_{k=-\infty}^{\infty} d_k r^k \ .$$

No other number system of the ancient times lends itself to such an elegant general format as the above. Gottfried Wilhelm von Leibniz considered the case of $r = 2$, modern binary number system in 1689 using zeros and ones and wrote an article called "Explication de l'Arithmétique Binaire" or "Explanation of the Binary Arithmetic" in 1703.

This may be called the VALUE FORMULA. This is in the form of a Sum of Products $d_k \bullet r^k$ which are obtained by multiplying the corresponding elements in the two arrays- one of weights and the other of digits. Notice that the number of symbols for the members of the basis set is limited to TEN in the INS and r in general. Imagine the problem of finding symbols for each of the numbers named in the list in Section 1.1.1. In the INS there is no such problem; number of any size can be written only in terms of the finite number of symbols equal to the radix. In all other ancient systems, the number of symbols goes on increasing.

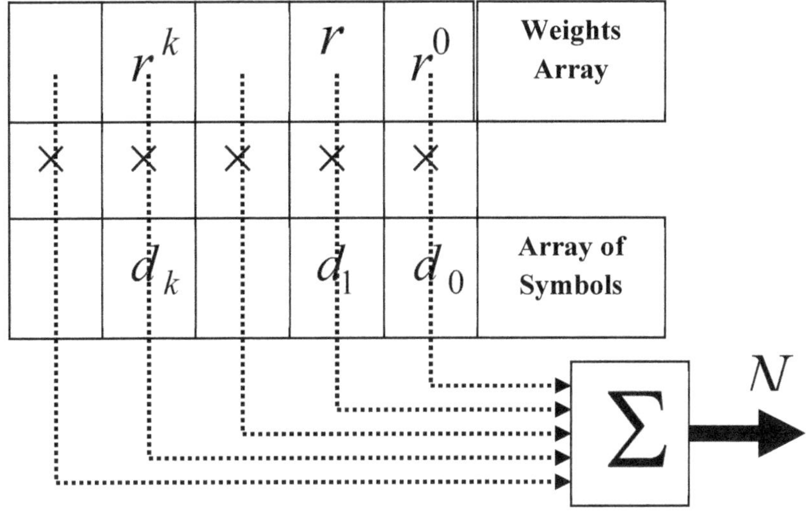

Figure 4.1 The arrow-pair in the INS: one array for representation in digits and the other for weights in powers of the radix

4.3   Pair Significance in Measure Determination- the Map and Scale Example

In the full REPRESENTATION of any entity, two arrays are implied. One denotes what is shown and the other an array which calibrates or scales the representation. This feature can be seen in other non-mathematical representations. For example, take the case of a map representing a geographical area shown in Figure 4.2 (a). From this representation we do not have any idea of the size of the island. To know about the size, we need a scale as shown in Figure 4.2 (b).

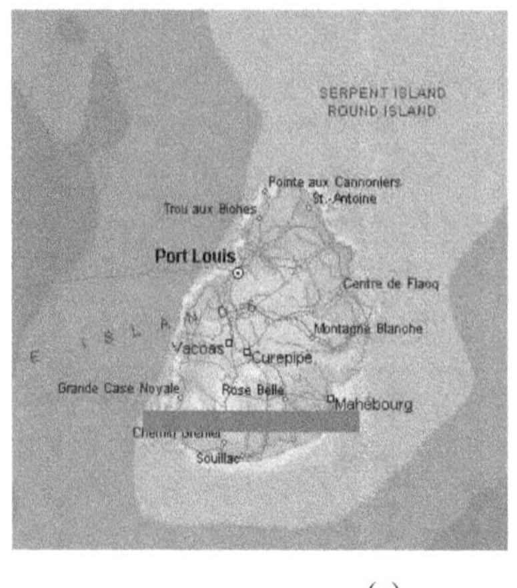

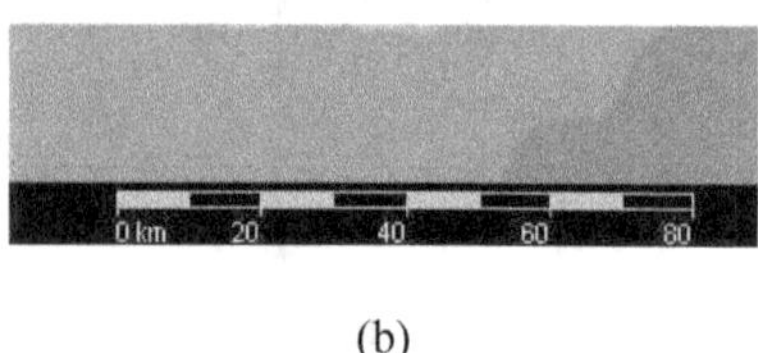

(a)   (b)

Figure 4.2 Map and scale example

With these two entities we will know the actual size of the real object in Figure 4.2 (a). In the same way if we have a number representation

101,

We cannot say what the number is unless we are told about the base or radix for this representation. If the base is 10 then it is understood as $10^2 \times 1 + 0 \times 10^1 + 10^0 \times 1$ which is 'one hundred and one'. But if the base is two, then the value of the same representation will be $2^2 \times 1 + 2^1 \times 0 + 2^0 \times 1$ which is 'five'.

The INS value formula

$$N = \sum_{k=-\infty}^{\infty} d_k r^k,$$

may be represented in the forms shown in Figure 4.3. Notice the entities d, r, k in the expression for N.

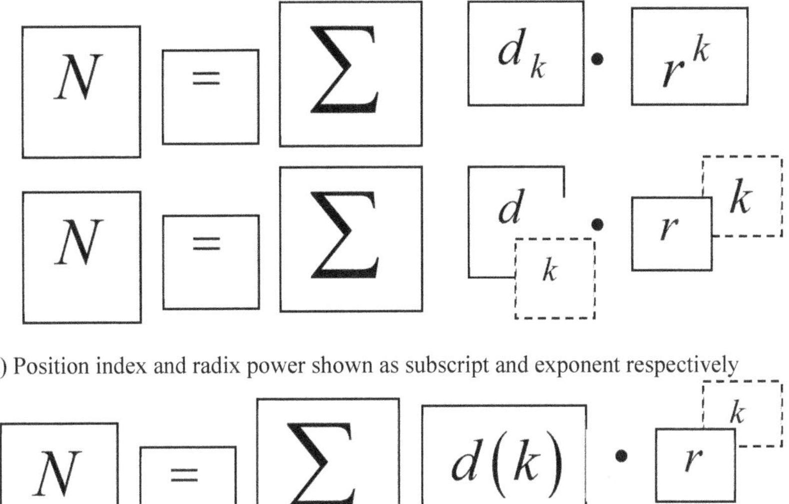

(a) Position index and radix power shown as subscript and exponent respectively

(b) Position index and radix power shown as argument and exponent respectively

Figure 4.3 Alternative placement of the position index in the INS format

In Figure 4.3(b) the subscript for d is shown as an argument in parentheses.

Evolution of Modern Forms of Representation of Quantities from the INS

When one notices the admiration of great minds in the West who say that the world has changed after the arrival of the INS, there must be deeper reasons beyond the widespread use in day-to-day life. The changes in the world of mathematics and the impact of the INS need to be investigated. As the author was unable to see any work in the literature, he was motivated to find the same by himself. Regarding the use of the name INS to refer to the number system under discussion this number system is sometimes referred to as Hindu-Arabic System but this is due to the important role played by the Arabs in disseminating it in the West; the works of the Arabs clearly attribute it only to the Hindus and always refer to it only as ' *tariq al-hindi* (the Indian way)'. The reasons behind the assertion that the field of mathematics was immensely influenced by the INS is that all the modern formulas for evaluation of various forms of mathematical representations appeared in the West only after the INS was accepted and put to use all over the world and none before the 13th century CE. Table 5.1 shows the timeline of significant developments in mathematical representation of various types of entities and formulas for evaluating them; they are centuries after the appearance of the INS in the West.

5.1 Scalar Constants

In general, a scalar constant is a number N and is expressible in the INS format on the basis of any positive integer $r, r \geq 2$ as

$$\sum_{-\infty}^{+\infty} d_k r^k \ .$$

wherein $d_k, k = 0, 2, ..., d_{k-1}$ are digits from the basis $\{0, 1, ..., (r-1)\}$ and r is a positive integer ≥ 2. The entire expression is in terms of positive integers $d_k, k = 0, 1, 2, ..., d_{k-1}$. Of course, k too is an integer including $k = 0$. This is the case of all positive integers occurring in the expression. The fractional part appears after a dot which marks the end of the integer part appears on the right with digits in decreasing order of weightage r^{-k}, $k = -1, -2, -3, \cdots$ and so on. Each part of the string of the digits may extend to infinity; the integer part towards left and the fractional part to the right. **Here two arrays are present: one is that of digits from the basis and the other is that of powers of the radix r. The VALUE of the number is in the form of a *sum of products* (SOP) or the *weighted sum* which are the key features of this evaluating formula. This structure of the EVALUATION FORMULA is what makes the INS as one of the most valuable contributions by the Indians to the world in general and of mathematics in particular.**

In what follows, the author's perception of the presence of the structure of the INS in a wide class of EVALUTION FORMULAS in modern mathematics will be outlined. **The evaluation formula of the INS will be seen to play the role similar to that of DNA in the evolution of other mathematical concepts.** This evolution will first be seen in the mathematical representation of a single entity such as a varying entity, a function of time $f(t)$ over a specified interval, which by itself is an array in time if sampled. But this representation does not need to be evaluated. However, there are other common forms of function representation and they can be derived as objects evolved from the INS by allowing the various components of the INS to take higher evolutionary forms of number such as ***real, imaginary, complex*** etc.

Table 5.1 Significant developments in the West to represent various entities in quantitative form following the arrival of the INS

CE	Event
1637	Rene Descartes proposed Cartesian coordinates
1689 (1703)	Gottfried Wilhelm von Leibniz conceived the modern binary number system in 1689 using zeros and ones and wrote an article called "Explication de l'Arithmétique Binaire" or "Explanation of the Binary Arithmetic" in 1703.
1731	Alexis Clairaut gives formula for distance in Cartesian coordinates and the concept of distance in Euclidian space
1820s	British mathematician George Green proposes Green's function
1850s	English mathematician James Sylvester, and Arthur Cayley developed the algebraic aspect of vectors and matrices
1898	The first usage of the concept of a vector space with an inner product is due to Giuseppe Peano
.1800	Convolution was first mentioned in a book by the French mathematician Sylvestre François Lacroix (1765–1843).
1814	Pierre Simon Laplace formalized the concept of expected value which was useful in interpreting probabilistic entities.
1821–1822	Convolution appeared in the famous work *Théorie Analytique de la Chaleur* by Jean Baptiste Joseph Fourier (1768–1830)
1888	Francis Galton first wrote about correlation
1965	Lotfi Zadeh introduced the concept of fuzzy quantities and linguistic variables. In the subsequent years M. Sugeno and E.H. Mamdani proposed rules for defuzzification of fuzzy quantities which are based on weighted average and centroid concepts. These are arithmetic weighted average operations used in estimating the expected value of a random number. During the early years of developments in Fuzzy systems, the fuzzy school insisted that fuzzy quantities must be distinguished by their max min operations and not by any other having arithmetic calculations. The present author pointed to their self contradicting use of weighted average or centroid distance in defuzzification and suggested a method that is devoid of such a contradiction (Rao and Rutherford, 1980). Nevertheless the use of arithmetic in fuzzy inference has been carried out obviously for its convenience. The author's esteemed friend Professor Yakov Tzypkin even said that what all you can do with the fuzzy formalism, can be accomplished in a probabilistic setting.

The influence of the INS that became commonplace may not be easy to prove by evidence of references and acknowledgements by the originators of the modern concepts; after centuries, popular ideas are not referenced to in literature but taken for granted. A very popular device or method invented long ago which became commonplace does not receive citations as we see today. This is due to the fact that the human mindset becomes framed in the structure of a concept that is commonplace. In other words, the human mind comes under seizure of the concept and operates without being conscious of its origin. When we use a wheel or fire, which are great human inventions or discoveries, we never think of the first human who invented or discovered it. Also, the names of originators in such cases are also not known even today. In the case of the INS the origin is widely acknowledged to be Indian but due to the extensive regular use in our daily lives, further details are seldom sought by any user unless his work is especially on the topic of number systems as is the case with the present work.

If the above statements sound too much as a claim that the mathematical works were inspired by the INS, the author would put the situation in the following way. The invention of the INS seems to have anticipated the potential of the concept behind its format in driving mathematical developments in quantity modeling and evaluation over the centuries to come

Although in the systems used by some other communities in the ancient times numbers were reckoned in tens, they were not provided to the world in a form as perfect as that of the INS. The Indian mathematicians described the INS with such perfect structure that upon dissemination into the west through the Arab scholars, there was no room for its further improvement; it fixed the mindset of the western and world mathematicians with its format and the format of the INS is shown here to be the basic *genetic element* in the evolution of more advanced mathematical concepts that we use today. Before the arrival of the INS, there was not a significant mathematical development in the West for nearly seven centuries.

It is the conviction of the author that the observations, arguments and comments made from here onwards are his own as he is unaware of any work that reported on observations on the INS for its impact in detail on the developments in mathematics as presented here.

5.2   Constant Vector

Consider a vector $\mathbf{x}$ in n-dimensional Euclidean space $\mathbb{R}^n$

$$x = [x_1, x_2, \cdots x_n],$$

where x_k is the component in the k-th dimension. If we wish to know how large this vector is, one single VALUE called the MAGNITUDE has to be calculated. For this, we treat the vector as an array as in the INS. However, it has components which are all REAL. To apply the SOP formula, we need another vector in the spirit of weighting array.

In the INS all entities in both the arrays are positive (integers). In the case of a vector some components of the vector can be negative. This negativity does not mean that that component has a different or negative effect. Negative components do not reduce energy; it remains the same irrespective of the sign of the component.

The energy concept will be clear if we consider an electrical heating coil of resistance ρ ohms. If an electric current is passed through it the coil is heated. The heat generated over a period of time denotes the energy converted from electricity to heat. The coil is heated over a period of time to some extent if a current of value c is passed through it. Even if we change to direction of the current such that the current is mathematically denoted as $-c$ in one direction reckoned as positive and the other as negative, the amount of heat generated will be the same. This suggests that in order to preserve the energy content in a vector component independent of its sign we must use as companion array same as the original vector for evaluation purposes. That is, a vector is weighted by itself. The 'product' now becomes a 'square' and even negative components show their energy content values as positive products. Then the sum of these products/squares represents the total energy content of the given vector. The SOP has now become 'Sum-of-Squares (SOS)'. Since we need the magnitude of the vector, we take the square root of this sum. We have thus arrived at a result which is the 'Root of Sum of Squares'. For an equivalent value of the electric current that can produce the same heating effect, the square root of this value is taken. When we deal with alternating current, we take the heating effect by taking the SUM as integral over the period and take the mean of the root over this period. This is the so called root mean square (RMS) value in alternating electrical systems. This happens to be the resultant of the two components denoted in a Phasor Diagram (vector diagram in this special context of electrical wave forms) in electrical systems. If we choose the same array for the purpose and apply the SOP/SOS formula we get a sum of all positive real quantities

$$SOP = \sum_{k=1}^{n} x_k^2 \ .$$

As the weight arrow is a copy of the original vector itself, this became a sum of squares. The square root of this becomes the magnitude of the vector which is

$$\|\mathbf{x}\| = \sqrt{\sum_{k=1}^{n} x_k^2} \ .$$

It is also called the norm of the vector $\mathbf{x}$. It is a measure of the size of the vector and in vector space the distance from the origin of the point $\mathbf{x}$. It may be viewed as the energetic value of the vector by virtue of the squaring operation on all the components. In the case of two dimensions the situation is shown in Figure 5.1(a).

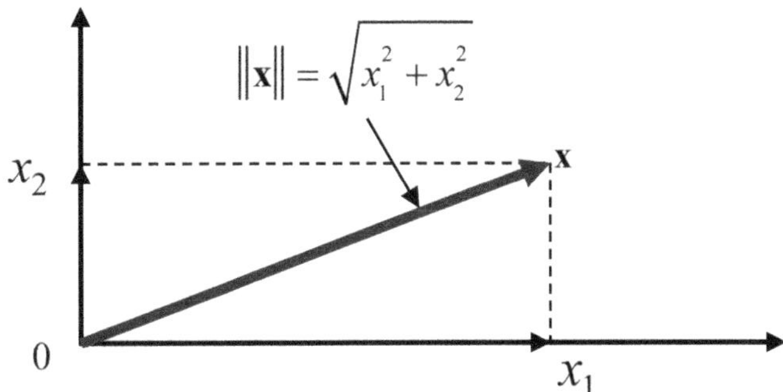

Figure 5.1(a) A two-dimensional vector in Euclidean space

Since we represented the vector in Euclidean space it is the denoted as the distance of the point $\mathbf{x}$ from the origin. Due to its equivalence with the value of the hypotenuse of a right-angled triangle having sides x_1 and x_2, it is also known as the Pythagorean distance. In general, in the n -dimensional Euclidean space $\mathbb{R}^n$, the norm of the vector is analogous to hypotenuse of a hyper right angled triangle.

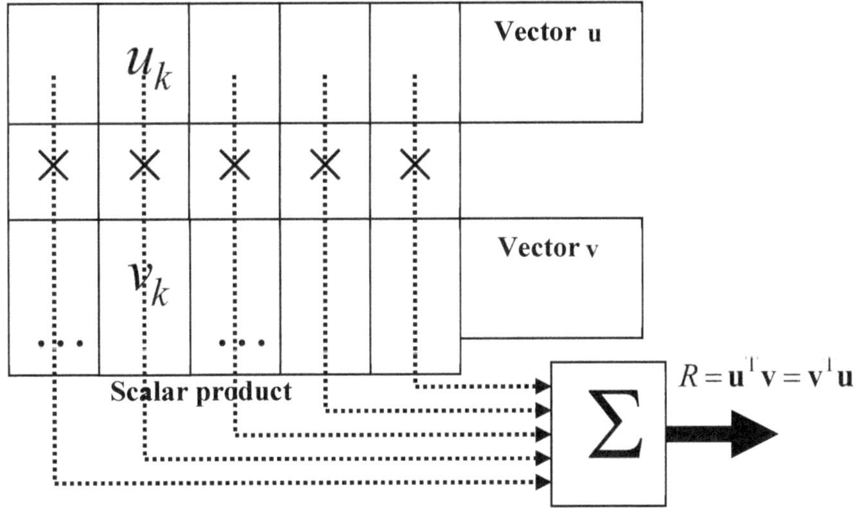

Figure 5.1(b) Two vectors as pairs in scalar product evaluation

Evaluation of the scalar product of two vectors is illustrated in Figure 5.1(b) which is akin to the evaluation of number from its representation.

5.3   Varying Quantities- Functions of Time

The numerical value of an entity may vary in time. In such a case the entity is denoted as a function of time $f(t)$. Such an entity may be graphically shown in Figure 5.2(a), the value of the entity f on the vertical axis as a dependent variable, and time on the horizontal axis as an independent variable.

A function of time is usually characterized in many ways just as we convert money into any currency or form for convenience of handling. Without any characterization a function of time $f(t)$ over a specified interval merely stands as a symbol denoting the function. One basic and crude way to characterize it is by a time series or sample values of the function as a numerical sequence over the interval. The number of samples depends on the sampling interval which if large leads to a small number of sample values, but if the sampling interval is small the number of samples becomes large and carries more information of the function; that is, the function is characterized more accurately.

Likewise, if the same function is characterized by an orthogonal series over the interval, the number of terms, like the number of samples, determines the accuracy of characterization. A function $f(t)$ over an interval is often expanded in Fourier series. We can consider orthogonal series expansions using other types of orthogonal basis functions. These functions may be seen as the so-called modulating functions (Shinbrot, 1957) used in the field of identification of continuous-time dynamic systems (Rao and Unbehauen, 2006). A modulating function is simply a function multiplying another- immediately suggesting the feature of the double array form akin to the product in the INS.

5.3.1   Power Series Representation of a Function of Time

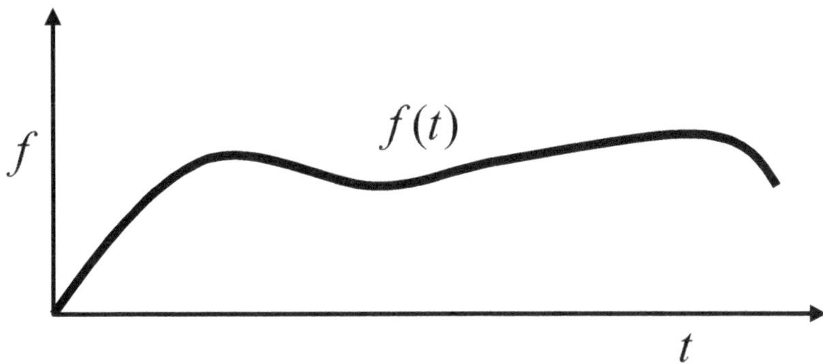

Figure 5.2(a) A function of time

In this representation time t denotes the position but is placed in brackets after f, as an alternative to the subscript notation in f_t . This change in the position indicator is to bring it into familiar format of a function of time.

$f(t)$ is written in a power series form as:

$$f(t) = \sum_{k=0} f_k t^k .$$

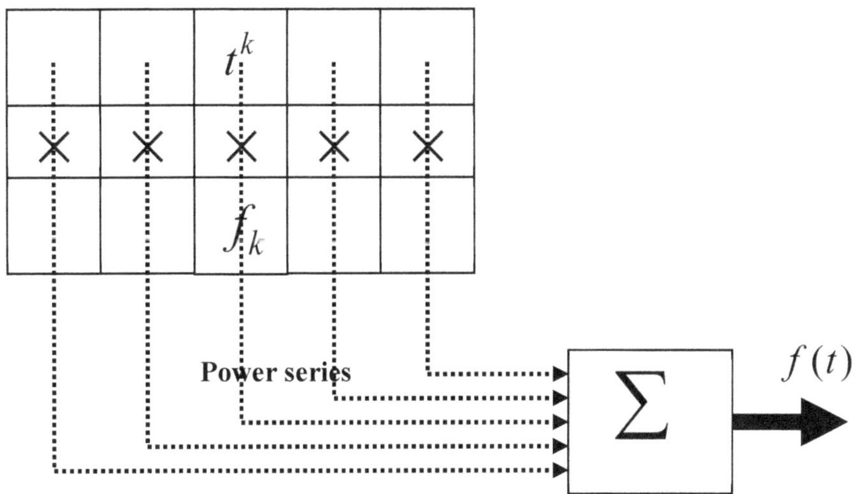

Figure 5.2(b) Power series representation of a function

This is shown in Figure 5.2(b) which is in the same format as that of the INS. Here f_k are called coefficients of the series expansion. The basic expression for a constant here takes a more general form with f_k and t having real values. The series representation of certain functions such as the exponential function and trigonometric functions appear only after the INS, especially in the Kerala School by Madhava and Parameswara in the 14th Century. The similarity of the above expansion with the INS is obvious. It appears as though the positive integer digits are allowed to take real values as coefficients in the sequence of weights - the radix r, a positive integer greater than 2 in the INS, is allowed here to take a more general form as the real variable t. Looking at the series the other way, basis functions $\left\{ t^k \right\}$ are polynomials of increasing degree. In the case of the exponential function the series representation is

$$e^t = 1 + t + \frac{t^2}{2} + \frac{t^3}{3!} + \cdots,$$

which can be written for its more general form

$$e^t = \frac{1}{0!} + \frac{t}{1!} + \frac{t^2}{2!} + \frac{t^3}{3!} + \cdots + \frac{t^k}{k!}.$$

Now let us probe the use of the **imaginary numbers** in the INS format by considering an imaginary radix $e^{j\Omega t}$.

A function of time defined over interval of time $\left[0, T \right]$ carries infinite amount of information in continuous time. If all the infinite set of points are used to denote two arrays one of continuous time and the other of the values of the function at the infinite number of instants of time, the problem becomes infinite dimensional. If, the function is represented in terms of n pulses, each of width ΔT as in Figure 5.3(a), the height of the pulse over the k-th subinterval being the average value of the function over that subinterval, a pair of arrays is created over the $[0,T]$. These pulses are block-pulse functions (Rao, 1983). The block pulse functions (BPF) are mutually orthogonal and also contiguous functions. The block pulse series $[f_0, f_1, ..., f_{n-1}]$ form a discrete array which may regarded as an n-vector in an n-dimensional Euclidian space $\mathbb{R}^n$.

The vector treatment previously presented applied here directly. As the subinterval is made infinitesimally small, the number of block pulses approaches infinity and the sampled sequence approaches the continuous time function $f(t)$ as an infinite dimensional vector.

We consider a function $f(t)$ over a defined interval which for our present purposes can be unity, resulting in a **normal interval** as it is called. Normalization will simplify the averaging process over the interval without introducing a scale factor.

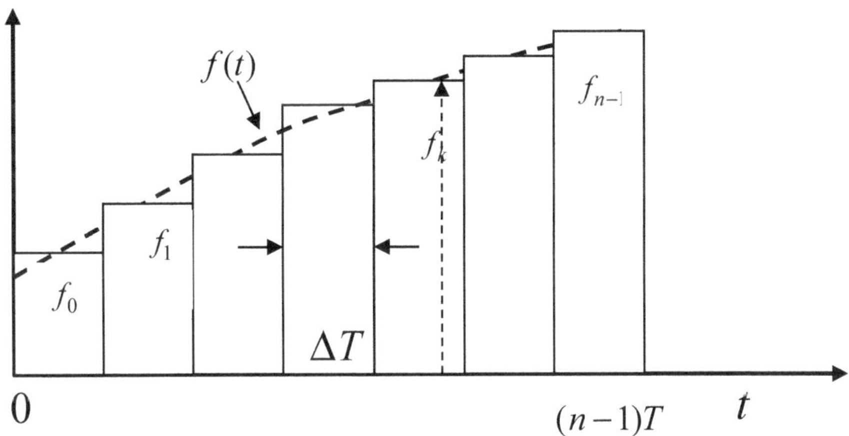

Figure 5.3(a) A function of time sampled over the period of its definition

To apply the SOP formula, we need another vector in the spirit of weighting array as in the case of vectors. If we choose the same function $f(t)$ as shown in Figure 5.3(b) for the purpose and apply the SOP formula we get the integral

$$SOP = \int_0^T f^2(t)dt \ .$$

The square root of this becomes the magnitude of this

$$\|f\| = \sqrt{\int_0^T f^2(t)dt} \ ,$$

which is the **norm or RMS value** or the norm of the function in the function space. It gives the size of the function analogous to the magnitude of a vector, the resultant of all the components. **To normalize a function we divide it by its norm.** A normalized function has unit energy. For our main discussion on functions, we consider functions over normal interval containing a unit of energy. Such normalization enables us to focus on the features of similarity we look for without any distraction created by scale factors.

In this way, the norms in vector and function spaces are seen to result from the weighted sum formula, involving the array pair which is the inherent measure feature of the INS.

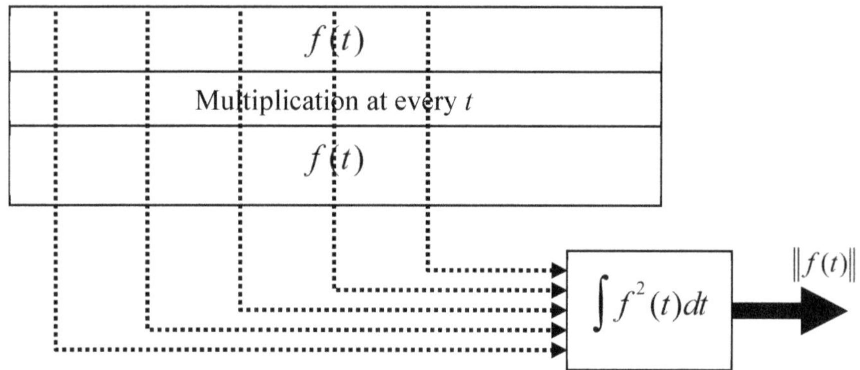

Figure 5.3(b) Evaluation of the norm of a function

5.4   Fourier Series Representation of a Periodic Function

In the special case if a function $f(t)$ that is periodic over an interval $[0,T]$ is multiplied by a function $\exp(jn\Omega t)$ or $\exp(jn\Omega).t$, $n=0,1,2,...$ where $\Omega=1/T$ the result

$$f(t) = \sum_{k=0}^{\infty} f_k \left[\exp(j\Omega t)\right]^k = \sum_{k=0}^{\infty} f_k \exp(jk\Omega t)$$

becomes the Fourier series representation of the function $f(t)$ with fundamental frequency Ω . *Notice that this is the case of the expression for the general number system format wherein the radix or base is changed into an imaginary form.* The sequence $f_n, \{n=0,1,2\cdots,\infty\}$ is the discrete Fourier spectrum of the function. It is analogous to the sequence of digits in the INS representation of a number. If we speak of the bandwidth, it corresponds to the WORD LENGTH of the number in the INS. In this case it is the largest value of k used in the expansion.

The sequence $f_k, \{k=0,1,2\cdots,\infty\}$ is the Fourier spectrum of the function Ω is the frequency of the fundamental and the various components are called higher harmonics. The case of $k=0$, meaning zero frequency or the DC (the familiar direct current component in terms of electrical signals) component happens to be the average value of the function over the interval because it follows from

$$f_0 = \int_0^T f(t)\exp(0)\, dt .$$

A particular frequency component is selected and separated for evaluation as

$$f_k = \int_0^T f(t)\exp(jk\omega t)dt .$$

The sequence $f_k, \{k = 0,1,2\cdots,\infty\}$, the discrete Fourier spectrum, represents the function $f(t)$ and approximates better with increasing number of terms in the series. The sum of squares of the spectral coefficients happens to be square of the function. The roots are the norms. Notice that the norm of a function can be determined by taking the square root of the sum of squares of the orthogonal components.

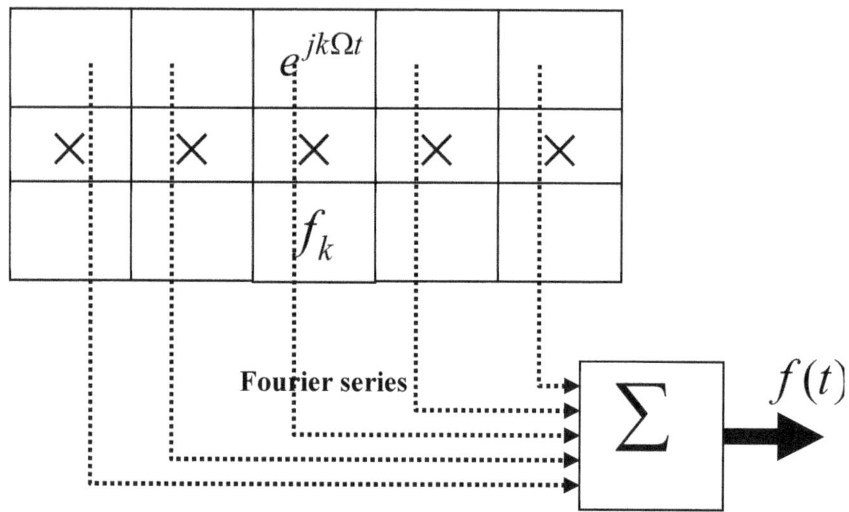

Figure 5.3(c) Fourier series representation of a periodic function

A series of orthonormal basis functions used in representing a given function is a complete set, each member carrying one unit of energy. The evaluation of the components by inner product is inherently a method of determining the distribution of the total energy of the function among its components. In fact, the sum of squares of all components in the expansion of a function on the basis of an orthonormal set $\{\varphi_k(t); k = 0,1,2,\cdots\infty\}$ is equal to the energy of the function $f(t)$. This is the meaning of the so called Parseval's equation

$$\sum_{0}^{\infty} f_k^2 = \int f^2(t)dt$$

5.5   Generalized Functions

The Dirac delta function, symbolically represented as $\delta(t)$, is an impulse function occurring at $t=0$. It is not a function in the ordinary sense - δ does not represent its magnitude at time t . It is a limiting pulse function whose height approaches infinity as its width is reduced to zero. **It is one place in the modern mathematical world where both zero and infinity occur in the same context**. The magnitude of the function is not the height of the impulse which actually is immeasurably large. Instead, the magnitude of the impulse is the **area under the impulse** in the limiting case. An impulse of unit area is called unit impulse. Thus the function is inherently defined in terms of AREA under the pulse that implies two dimensions- $height \times width$.

Unit impulse function is defined by the following integral

$$\int_{0}^{\infty} \delta(t)dt = 1 .$$

In general it is also given by the following integral form:

$$\int_{0}^{\infty} \varphi(t)\delta(t-\tau)dt = \varphi(\tau) .$$

Notice that in the integrand above there is no term in the form of a radix raised to a power, but in its place there is the shifted Dirac delta function. This is clearly a *natural positioning feature* which in other forms is seen as a positioning term $(\bullet)^{(\bullet)}$ which first has been seen in the INS. Thus, the presence of the INS feature here may again be noted.

One cannot directly measure an impulse since it destroys the measuring instrument due to its property of infinite shock. Therefore, it is scaled or multiplied by a *measuring function* $\varphi(t)$ and the product is integrated to get a value of $\varphi(\tau)$. If the impulse is of magnitude m the integral gives a value of $m\varphi(\tau)$. It must be noted that it is again an application of the formula for evaluation from the INS. The process is multiplication of a pair of time functions in the integrand followed by summation which is the SOP feature. Thus, the symbol $\delta(t)$ attains a value only in association with another time function and applying the SOP formula as an integral. In the first form in the integrand, we have to note that $\delta(t)$ has a companion – a measuring unit function $1(t)$ that is not written in the integrand; it is implied and latent.

In linear dynamic systems the system response to a unit impulse is the characteristic of the system – its basic property. An impulse should never be used in experiment for direct evaluation of the impulse response since the experiment will be a destructive shock on the system. Response for other forms of input can be obtained and from them the impulse response can be deduced.

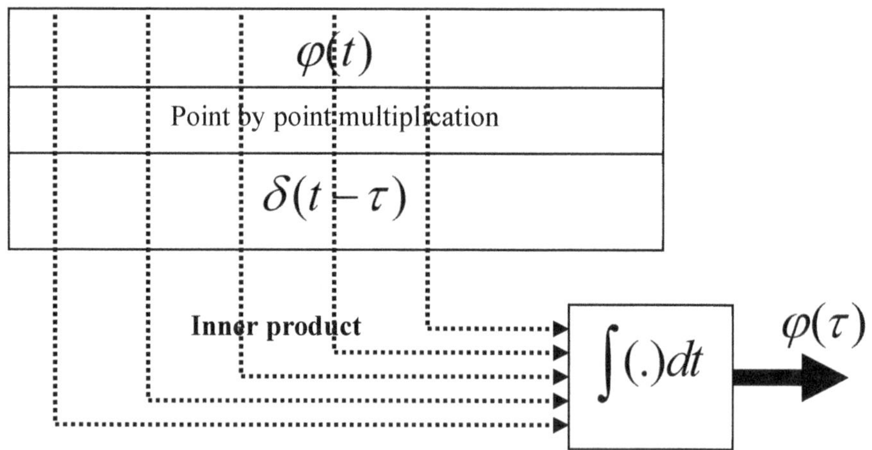

Figure 5.3(d) Evaluating a Dirac delta function - generalized function

5.6   Evolution of Integral Transforms: Transformation of Variable t to s

Now we will allow the entities in the INS structure to take the form of complex variables to see how integral transforms can be generated. If a given function $f(t)$ $t \in [0, \infty]$ is multiplied by a function in two dimensions $K(s,t)$ where s is a complex variable and integrate over time, the result is an integral transform of $f(t)$. The original function $f(t)$ is transformed into the s domain

$$F(s) = \int_{t=0}^{\infty} f(t)K(s,t)dt \ .$$

If the kernel $K(s,t)$ is chosen in the special form $\exp(-st) = \{\exp(-s)\}^t$ we have the familiar Laplace transform.

Notice that here the radix is *complex* in the variable s - an exponential function of s . The kernel denotes a mathematical operation for placement in position t . That is, $K(s,t)$ **has to be literally understood to mean 'at t '.** Recall that in the s domain the operator to position at a point of time T is well known as e^{-sT} . Thus $\exp(-st) = \{\exp(-s)\}^t$ mathematically defines the position which in this continuous time case is t, that is, the value of the function f at t which is written in the well-known form as $f(t)$. Here the position index is written in parentheses (.) adjacent to f as alternative to the subscript notation as f_t . Laplace transformation formula can be guessed by the properties required by $K(s,t)$ to facilitate conversion of derivatives and integrals into algebraic forms in the complex variable s . For conversion of differential equations into algebraic form we need two important results: For differentiation we need, say $\mathcal{L}(df / dt) = sF(s)$ and for integration $\mathcal{L}\left(\int f(t)dt\right) = \frac{1}{s}F(s)$.

It is the exponential function e^{-st} as the kernel $K(s,t)$ that retains its form both in differentiation and integration but scaled by s and $\frac{1}{s}$ respectively. Thus, with the said exponential function as the kernel, functions in time t are transformed into those of the complex 'frequency' thereby transforming the calculus of linear differential equations from real t into an algebra in the complex variable s that is easier to handle.

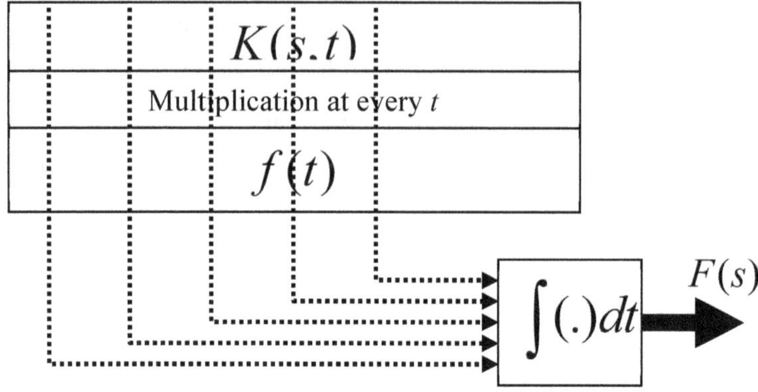

Figure 5.4(a) An integral transform

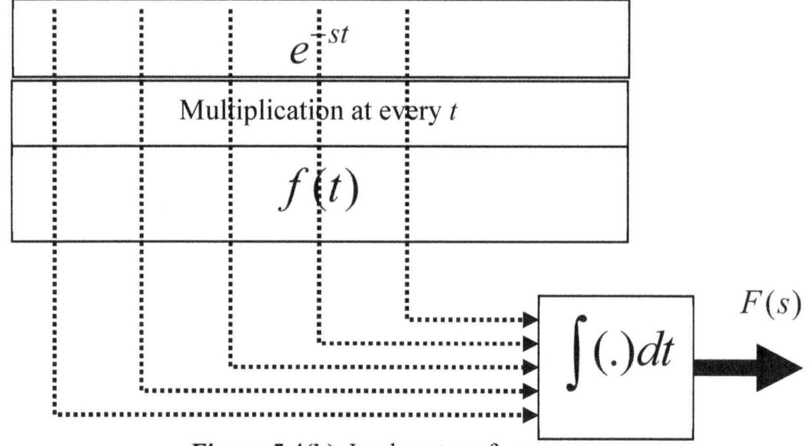

Figure 5.4(b) Laplace transform

A special case of the above is the Fourier transform wherein s is replaced by the frequency variable $j\omega$.

Another special case of the above is the z-transform of the sampled function $f(nT); n = 0, 1, 2 \cdots$.

$$F(z) = \sum_{n=0}^{\infty} f(nT)\left(z^{-1}\right)^n$$

Notice that here the place of the radix is occupied by the shift operator z^{-1}

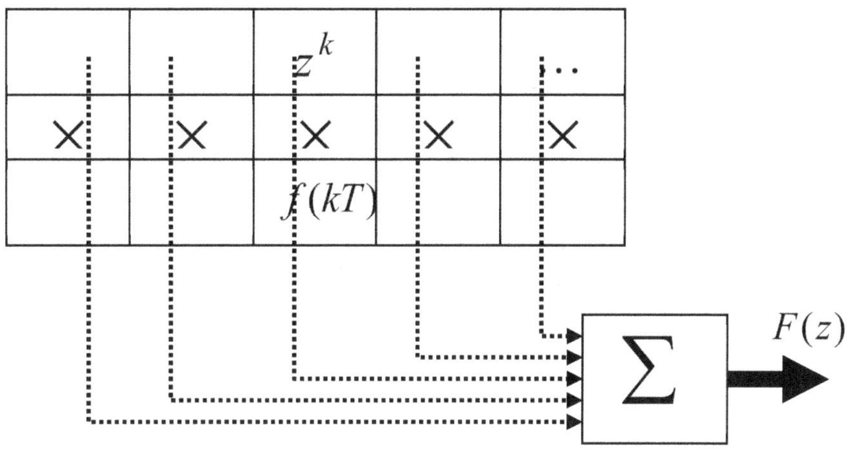

Figure 5.4(c) z-transform

The underlying structure of these integral transforms and of the Fourier series representation of a function $f(t)$ are the same as of the INS. In the INS a number is reckoned in bunches of TENS- including TENS of TENS and so on $(10, 10^2, \ldots)$; in the discrete Fourier series, a function is reckoned in sinusoidal functions and their terms of higher order obtained by successive powers, and in the Laplace transform the function is reckoned in terms of the exponential function (e^{-s}) and its powers of t giving rise to the notion of complex frequency (s).

5.7   Mellin Transform

The Mellin transform of a function $f(t)$ is given by

$$F(m) = \int_0^\infty f(t)\, t^{m-1} dt$$

Expression	Number in INS	Power series expansion	Discrete Fourier series of a Periodic function	Generalized function	Laplace transform	Fourier transform	z-transform	Mellin transform
Radix	Positive integer $r \geq 2$	Real, t	Imaginary $\exp(j\Omega t)$	Dirac delta function $\delta(t)$	Complex $\exp(-s)$	Imaginary $\exp(-j\omega)$	Shift operator z^{-1}	Real t
Coefficient	Positive integer $0 \leq d_k \leq r-1$	Real f_k	Spectral component f_k	Generalized function	Real $f(t)$	Real $f(t)$	Real f_k	Real $f(t)$
Position index	Integer k	Integer k	Integer k	Real τ	Real, $t \in [0,\infty]$ continuous	Real, $t \in [0,\infty]$ continuous	Positive integer k	t
Position value	r^k	t^k	$\exp(j\Omega t)^k = \exp(jk\Omega t)$	$\delta(t-\tau)$	$[\exp(-s)]^t$ $= \exp(-st)$	$[\exp(-j\omega)]^t$ $= \exp(-j\omega t)$	$\left(z^{-1}\right)^k = z^{-k}$	t^{m-1}
Remarks	Word length= number of digits in the number expression		Discrete fourier spectrum	Natural positioning along t due to the shift	In general (α^{-s}) is also possible but with e^{-s} the transform is normalized	Continuous spectrum	Positioning natural due to the shift	

The similarity of the above with the INS is self-explanatory. Table 5.2 shows all these representations as the results of evolution of the radix from positive integer form to other general forms. As the radix takes different forms from positive integer- real- imaginary-complex forms, and the coefficient from positive integer to real forms, and the exponent and the position index from integer form to a real form, a variety of species of mathematical expressions for representation of entities emerges from the basic INS.

5.8   The Reverse Process: Tuning, Determination of the Parameters in the Representation Forms

The process of obtaining the parameters of a representation of a number or a function is practically important. The various cases of representation and the process of determining the parameters are given below

5.8.1 Number System

For a number N in the number system on a base r the individual digits are given by the following formula

$$d_k = \left\lfloor \frac{N}{r^k} \right\rfloor - \left\lfloor \frac{N}{r^{k+1}} \right\rfloor r,$$

in which the symbol $\lfloor \quad \rfloor$ denotes 'integer part of'.

5.8.2   Power Series

In the power series representation of a function $f(t)$, the coefficients in the power series are given by

$$f_k = \frac{1}{k!}\frac{d^k}{dt^k}f(0),$$

which is from McClaurin series.

5.8.3 Discrete Fourier Series of a Periodic Function

$$f_k = \frac{\Omega}{2\pi}\int\limits_{-\pi/\Omega}^{\pi/\Omega} f(t)e^{-jk\Omega t}dt .$$

The coefficients f_k in the case of given function are the values of the frequency components, Obtaining a component f_k is to pick it up from a mix. The selection is also called TUNING. When we tune a system for the k-th harmonic component, it is sensed with maximum sensitivity and all other components are filtered out. In radio networks a circuit is tuned to the resonant frequency to pick up signal at the required frequency.

It is therefore interesting to see that all the above processes have the same objective which is selective determination.

In view of the above it is also interesting to note that BANDWIDTH of a signal is akin to the WORD LENGTH in a number. This feature is dependent on the basis of representation.

For example the representation of number 1/3 in the decimal system happens to be

$$0.3333333...$$

It is like a very wide bandwidth if you view each digit as a spectral component. If you change the basis of the number system to 3, which gives the **ternary representation**, the same number appears as

$$0.1$$

becoming band limited. Thus, basis is a fundamental set of entities chosen to express any other arbitrary entity. In the decimal number system this is the set of ten numbers including zero- **{0,1,2,3,4,5,6,7,8,9}.** In the binary system of numbers it is **{0,1},** In Fourier series, the basis is a set of **sinusoidal functions** and in power series it is $\{t^k , k = 0,1,2,3,...\}$. **Thus the terms basis, spectrum, tuning, band width, etc. which we name in the case of sinusoidal functions, appear to belong to more general aspects in the representations of other entities. The INS has opened up this view by being the fundamental form of a mathematical representation.**

5.8.4 Generalized Functions

The generalized function is given by $\varphi(\tau)$

$$\phi(\tau) = \int_{-\infty}^{\infty} \varphi(t)\delta(t-\tau)dt .$$

5.9   The Reverse Process- Inversion of Transforms

In the case of functions which have been transformed into another domain, the original functions in the native domain are obtained by the following relations:

5.9.1 Inverse Laplace Transform

The function $f(t)$ is obtained from its Laplace transform $F(s)$ using the inverse formula

$$f(t) = \frac{1}{2\pi \, \mathrm{j}} \oint F(s) \, (e^t)^s \, ds .$$

5.9.2 Fourier Transform

The function $f(t)$ is obtained from its Fourier transform $F(\omega)$ using the inverse formula

$$f(t) = \frac{1}{2\pi} \oint F(\omega)\,(e^{j\omega t})\,d\omega.$$

5.9.3 z-Transform

The function $f(nT)$ is obtained from its z-transform transform $F(z)$ using the inverse formula

$$f_k = \frac{1}{2\pi j} \oint F(z) z^{k-1}.$$

5.9.4 Mellin Transform

The function $f(t)$ is obtained from its Laplace transform $F(m)$ using the inverse formula

$$f(t) = \frac{1}{2\pi} \oint F(m) t^m dm.$$

In general, the inverse transforms for $F(s) = \int f(t) K(s,t) dt$ are in the form

$$f(t) = \oint F(s) K^{-1}(s,t) ds,$$

where, $K^{-1}(s,t)$ is the inverse kernel. These are summarized in Table 5.3.

Table 5.3 The relations in the process of reversion to the original form of an entity

Represented candidate	Representation	Parameters of represe-ntation	The process of finding the parameters and inverse transforms
Number N	$N = \sum_{-\infty}^{+\infty} d_k r^k$	d_k	$d_k = \left\| \dfrac{N}{r^k} - \dfrac{N}{r^{k+1}} \right\| r$
Function $f(t)$	$f(t) = \sum_{k=0} f_k t^k$	f_k	$f_k = \dfrac{1}{k!}\dfrac{d^k}{dt^k} f(0)$

Contd...

Represented candidate	Representation	Parameters of represe-ntation	The process of finding the parameters and inverse transforms
Discrete Fourier Series of a Periodic function	$f(t) = \sum\limits_{k=0}^{\infty} f_k e^{jk\Omega t} dt$	f_k	$f_k = \dfrac{\Omega}{2\pi} \int\limits_{-\pi/\Omega}^{\pi/\Omega} f(t) e^{-jk\Omega t} dt$
Laplace Transform	$F(s) = \int\limits_{0}^{\infty} f(t) e^{-st} dt$	$f(t)$	$f(t) = \dfrac{1}{2\pi j} \oint F(s)\, e^{st} ds$
Fourier Transform	$F(\omega) = \int\limits_{0}^{\infty} f(t) e^{-j\omega t} dt$	$f(t)$	$f(t) = \dfrac{1}{2\pi} \int\limits_{-\infty}^{\infty} F(\omega) e^{j\omega t} d\omega$
z-Transform	$F(z) = \sum\limits_{n=0}^{\infty} f(nT)\left(z^{-1}\right)^{n}$	$f(nT)$	$f_k = \dfrac{1}{2\pi j} \oint F(z) z^{k-1}$
Mellin Transform	$F(m) = \int\limits_{0}^{\infty} f(t) t^{m-1} dt$	$f(t)$	$f(t) = \dfrac{1}{2\pi} \int\limits_{0}^{\infty} F(m) t^{m} dm$
General Integral Transform	$F(s) = \int f(t) K(s,t) dt$	$f(t)$	$f(t) = \int F(s) K^{-1}(s,t) ds$ $K^{-1}(s,t)$ is the inverse kernel

5.10   Broader Meaning of Certain Terms related to Spectral Representation

Basis: Basis is set of entities in terms of which an entity is expressed. The following examples share the feature of a basis in some way or other although they are not strictly in a mathematical setting.

Wealth in the ancient times was reckoned in terms of livestock numbers.

It is sometimes expressed in terms of the units of land one possesses.

In expressing the amount of facility in an apartment we use the basis as bedroom, kitchen, hall, etc. e.g. 2BHK, 3BHK, etc.

In our times it is expressed in terms of money such as dollars, pounds, euros, rupees etc.

The numbers of these units of currency are made up of notes and coins of different denominations such as 100s, 20s, 10, and so on. A number by itself is expressed on the basis of a base or radix which in our daily life is the set [0, 1, 2, 3, 4, 5, 6, 7, 8, 9] which is the decimal system.

In the case of currency amounts less than one unit, we see coinage of smaller denominations such as cents, pence, etc. These smaller units are minted to represent certain numbers of the smallest unit. For instance, in the case of cents in the USA a dime, a quarter, are also minted to denote 10 and 25 cents. Many decades ago coinage of the United Kingdom, below one pound was akin to binary system which has been changed into decimal system for compatibility with calculations which are in the decimal system.

Electronic computers on the other hand work on the basis of the binary system due to its simplicity and compatibility with the ON and OFF states of an electrical circuit – denoted respectively as 1 and 0.

In number systems the basis can be any positive integer greater than or equal to 2. In the basis, each of the elements has to be denoted by a symbol. In the binary system the basis is [0,1] and in the decimal system [0,1,2,3,4,5,6,7,8,9]. Hexadecimal system is also used sometimes and in this symbols of the decimal set and additional symbols for eleven, twelve, ... fifteen, have to be used. These additional symbols can be A, B, C, D, and E.That is the basis in a hexadecimal system will be[0,1,2,3,4,5,6,7,8,9,A,B,C,D,E]. [0,1,2,3,4,5,6,7,8,9].

In power series representation of a function the content of the function is in terms of polynomials of degree 0, 1, 2, etc.

There exist sets of orthogonal polynomials which are also used as basis for expansions of given functions. Examples are Jacobi, Legendre, Laguerre, Chebyshev, Bessel functions. These are infinite countable sets.

There exist piecewise constant orthogonal functions such as Block Pulse functions, Walsh Functions, Haar Functions, etc. to effectively represent piecewise constant functions (Rao, 1983). Likewise, one may view wavelets which generally come under the modulating functions. Haar functions may also be used to get wavelets.

For function expansion, the chosen basis set must be complete. Completeness ensures both even and odd components in the set to properly approximate a given function.

When functions over an interval are expressed in terms of orthogonal basis functions, the orthogonality/ orthonormality of the basis functions makes each component in the expansion orthogonal to the rest. This property is desirable since the disturbances or errors in any component do not affect the rest of the components. Furthermore, additional higher order components can be included and this addition does not affect the existing components.

In the spectral representation of a signal the information content is reckoned in terms of the values of the components at various frequencies.

Sequence of Coefficients in a Representation: This is a sequence of numbers denoting the amount or value of the respective components. This sequence is referred to in different contexts according to its nature

In the decimal number system the number 1977 implies that there is 1 thousand (10^3), nine hundreds (10^2 s), seven tens (10s) and seven units (1s).

The binary representation 101 of a number means there is one four, no twos, and one unit in the number- it is five.

In the Fourier series representation of a period function over its period, the sequence of coefficients is called the Fourier Spectrum. It shows the information content in the function as components at different frequencies. Likewise in other orthogonal series representations, the sequence denotes the coefficients of the components arranged in order.

All these share the same mathematical status. For example, when we look at a decimal number it is in the same sense of a spectrum- each digit indicating the strength of that component.

Width of a sequence: This is now seen to bear different names in different situations, bandwidth in signals, word length in numbers, series length in power series expansions etc.

Tuning: This is in general the process of **selective determination** of a particular element in the set of coefficients.

5.11 Product-Convolution Duality in the INS Structure

Multiplying two numbers in the INS format can also be obtained by convolution of the arrays. This **product-convolution duality** is well known in the study of signals and their transformed versions such as the Laplace transform.

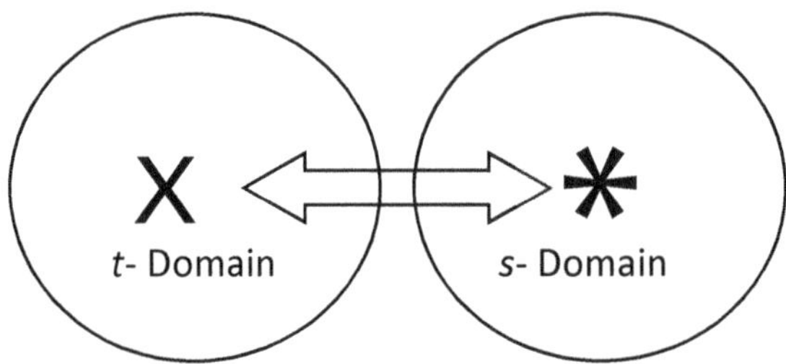

Figure 5.5 Product-convolution duality

Table 5.4 Product and convolution in the domains of s and t

Product	Convolution
$H(s) = F(s) \times G(s)$	$\left[h(t) = f(t) * g(t)\right] = h(t) = \int\limits_{0}^{\infty} f(\tau)g(t-\tau)$
$h(t) = f(t) \times g(t)$	$\left[H(s) = F(s) * G(s)\right] = \oint F(s-\alpha)G(\alpha)d\alpha$

Product by multiplying and collecting coefficients of equal powers of r

$$M = m_2 r^2 + m_1 r + m_0$$
$$N = n_2 r^2 + n_1 r + n_0$$

$$M.N = (m_2 n_2)r^4 + (m_1 n_2 + m_2 n_1)r^3$$
$$+ (m_0 n_2 + m_1 n_1 + m_2 n_0)r^2 + (m_0 n_1 + m_1 n_0)r + m_0 n_0$$
$$= \sum_{j=-(k-1)}^{k-1} \sum_{i=0}^{k-1} m_i r^i n_{i-j} r^{i-j} \quad , k=3$$

Product by applying the convolution formula

$M.N = \displaystyle\sum_{j=-(k-1)}^{k-1} \sum_{i=0}^{k-1} m_i r^i n_{i-j} r^{i-j} \quad , k=3$ as detailed in Table 5.5. In fact, we

multiply numbers in this way. Convolution is the process of multiplying two arrays with one of them shifted repeatedly by varying the shift over the range of the arrays and adding the results.

Table 5.5 Stages in the shift-multiply-add operations in digital convolution for multiplication of two numbers

		m_2	m_1	m_0			
n_2	n_1	n_0					
		$m_2 n_0 r^2$					$j = -2$
		m_2	m_1	m_0			
	n_2	n_1	n_0				
		$m_2 n_1 r^3$	$m_1 n_0 r$				$j = -1$
		m_2	m_1	m_0			
		n_2	n_1	n_0			
		$m_2 n_2 r^4$	$m_1 n_1 r^2$	$m_0 n_0$			$j = 0$
		m_2	m_1	m_0			
			n_2	n_1	n_0		
			$m_1 n_2 r^3$	$m_0 n_1 r$			$j = 1$
		m_2	m_1	m_0			
				n_2	n_1	n_0	
				$m_0 n_2 r^2$			$j = 2$
$m_2 n_2 r^4 + (m_1 n_2 + m_2 n_1) r^3 + (m_0 n_2 + m_1 n_1 + m_2 n_0) r^2 + (m_0 n_1 + m_1 n_0) r + m_0 n_0$							

Example 5.1: For simplicity let us calculate the product of SEVEN and THREE and work on the binary representation and convolution. With Seven (0111) and Three (011) let us apply the convolution formula (Shift and add only; then 11 gives rise to one row 111 and the next row 1110. The last 0 in the second row is not shown in Table 5.6 since it is implied. As we are so used to the shifted multiplication and addition repeated over the range as above, we usually do not recognize the process as discrete convolution. n order to get the result of multiplication *without convolution* is by the brute-force method, counting the number of cells formed by Table 5.7 of rows with dimensions of M and N as shown below in the case of THREE X SEVEN = TWENTYONE:

Table 5.6 Product of SEVEN and THREE expressed in binary system

		1	1	1
	1	1	1	
	1	**1**	**0**	**1**

Table 5.7 Multiplication by brute-force method of listing M, N times and counting

	M = Seven						
	X	X	X	X	X	X	X
N = Three	X	X	X	X	X	X	X
	X	X	X	X	X	X	X

Remarks 5.1: In the above case, we sought the product of two numbers M and N, but we used a discrete convolution method as though we were seeking the response of a discrete-time dynamical system. Why had we to use convolution? This can be explained in the following manner. The two numbers M and N were represented as horizontal sequences of digits (numbers limited to 0 to 9). Each location k in the INS can accommodate only radix powers 10^k at that location. Either M or N is fixed and the other shifted in steps until the end of the shifted string and the results of multiplication are all added. This is in fact

convolution. For instance, the product of the two digits at location k will produce results in (10^{2k})s which must be placed in the location $2k$. This spill over to a higher location is *carryover* in the language of arithmetic. The multiplication action at location k resulting in the higher order placement of the result is like an impulse applied at a given instant of time t to a dynamic system. The response of the system extends into future time. To understand the process let us consider multiplication of two numbers in the decimal system $M = 94; N = 58$ and place them in a table (Table 5.8) one above the other with digits in individual cells)

Table 5.8 Multiplication of 94 by 38 by convolution

Position index k	3	2	1	0	
Position value	10^3	10^2	10^1	10^0	
M	0	0	9	4	
$\times$			$\times$	$\times$	8x4=32 (shift=0)
N	0	0	5	8	
			3	2	The act of multiplication produced a result 32 with the digit 3 moved to next place. The result spreads to $k=1$
		7	2	0	8x90=720 (shift =0). The result spreads to $k=2$
		2	0	0	50x4=200 (shift=1). The result spreads to $k=2$
	4	5	0	0	50x90=450 (shift=1). The result spreads to $k=3$
	5	4	5	2	The result of addition

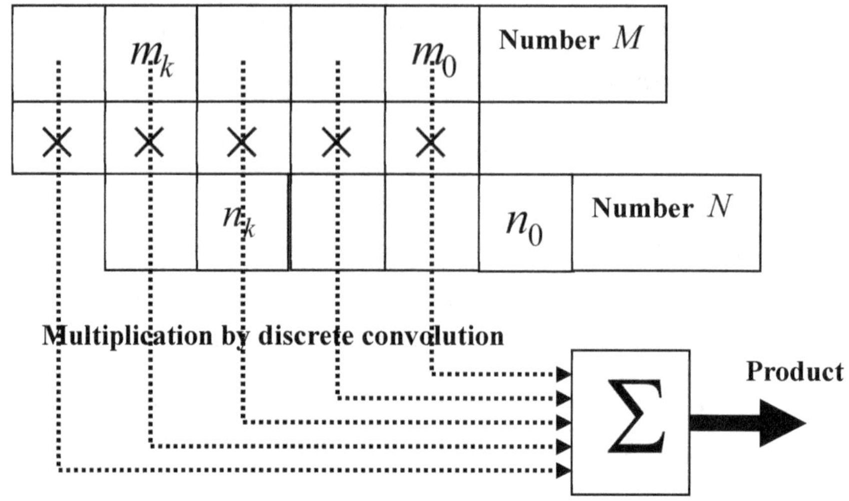

Figure 5.6 Multiplication of two numbers by digital convolution

According to tradition in the above k increases from right to left in steps of one. It is schematically shown in Figure 5.6.

Now consider a dynamic system shown in Figure 5.7.

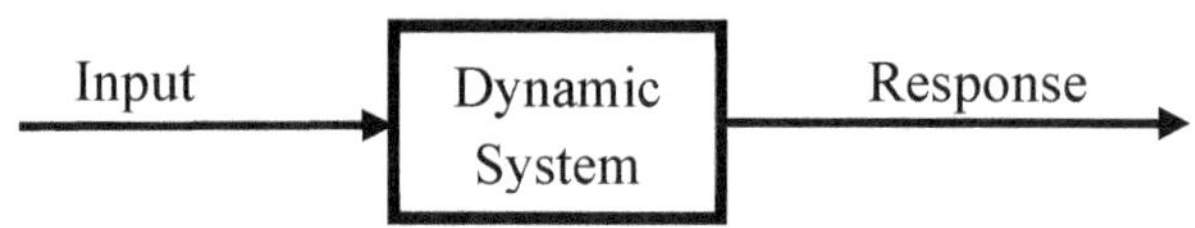

(a). A linear dynamic system

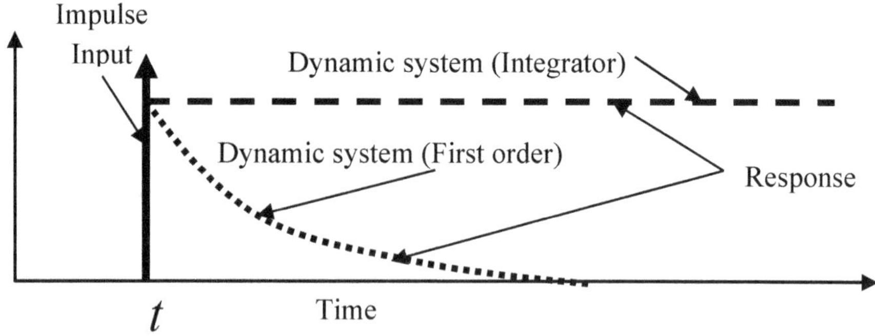

(b). Impulse response curves of simple linear dynamic systems

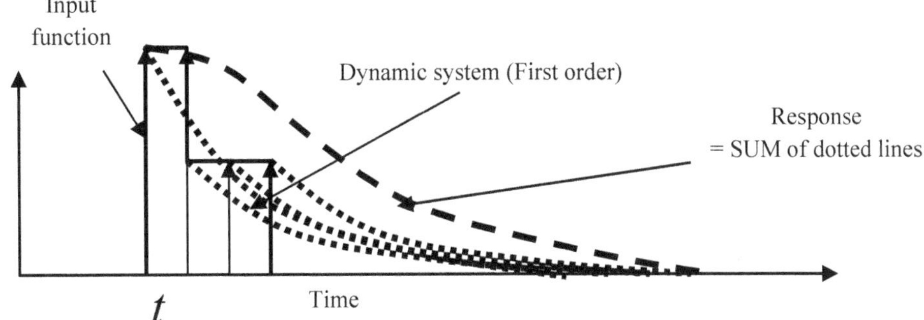

(c). A dynamic system response evaluation using impulse response in convolution with the input signal

Figure 5.7 A linear dynamic system and its response functions

According to tradition here increasing time is shown to the right. Notice how the response extends into the future

As in the case of numerical multiplication the input function and the impulse response function of the linear dynamic system can be interchanged but the response will be the same like the result of multiplication of M with N or vice versa.

Impulse response functions due to inputs applied at varying time τ (shift of the horizontal time axis) from 0 onwards are all added up to yield the total effect at any instant of time. Precisely this process is performed when we use the shift-multiply operations and add up the results in arithmetic. Thus, in multiplying two numbers expressed in the ten-based INS or on any other system with a different base r, we use digital convolution without being conscious of it. We become familiar with the term *convolution* only when we study liner dynamic systems for their time response. While the shift-multiply and add process is learnt very early in school, we learn about convolution only in higher education involving dynamic systems. The mathematics of linear dynamic systems and the concept of convolution were developed centuries after the arithmetic of the INS was introduced first by the Indian mathematician Brahmagupta. This arithmetic was explained by Fibonacci in his *Liber Abaci* in 1202. The study of linear dynamic systems entered into university curricula after World War II and gradually became widespread.

The similarity between the dynamic system response evaluation and multiplication of two numbers in the INS format is noteworthy. More about the convolution in linear dynamic systems will be discussed in a later section.

5.12   Random Quantities

A discrete random quantity is represented as $x_i(p_i)$, where x_i is the particular outcome and p_i is its probability. Then, its EXPECTED VALUE

$$\mu = E(X) = \sum_{1}^{n} x_i p_i \,,$$

which is the SOP formula as in the INS. In the arrays $\{x_i\}$ and $\{p_i\}$ the corresponding elements are multiplied and added to get the **EXPECTED VALUE.**

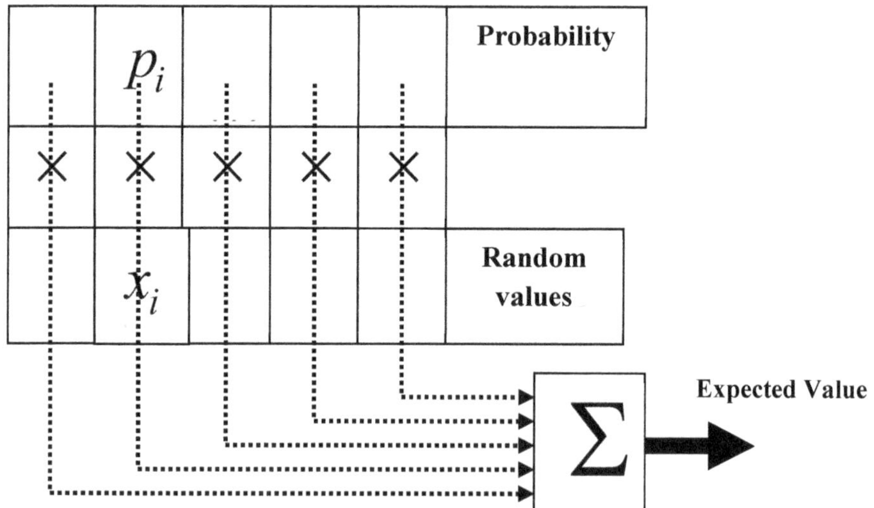

Figure 5.8 Evaluation of expected value of a random quantity

5.13   Fuzzy Sets

A fuzzy set S is characterized by a set of values, each associated with a membership value $\mu \in [0,1]$ shown in Figure 5.9(a). The very definition is in the form of the INS. The support values have corresponding values of membership in the set which are in the spirit of weights for the values of support. The membership is distinguished from similar looking probability function by the fact that it is not randomness but vagueness in the fuzzy case. Some examples of fuzzy variables are: SHORT, TALL, MEDIUM; YOUNG, OLD; LARGE, SMALL, etc. These are linguistic descriptions of some entities. On the other hand, we may also have fuzzy numerical variables such as ~3, ≤ 5 etc. Some of these fuzzy quantities are illustrated in Figure 5.9(b). Combinatorial rules for fuzzy sets are defined with 'max' ($\vee$) and 'min'($\wedge$) operations' denoting **union** and **intersection** of the sets respectively. Thus the combination of two fuzzy sets A and B having membership functions μ_A and μ_B respectively. The union $A \cup B$ is defined with a membership function $\mu_{A \cup B} = \mu_A \cup \mu_B$ and The intersection $A \cap B$ is defined with a membership function $\mu_{A \cap B} = \mu_A \cap \mu_B$. These are shown in Figure 5.9(c).

A set of statements such as 'If A Then B and if C then D' is combined in fuzzy logic to obtain the resulting fuzzy set for control action may be in the form shown in Figure 5.10. In order to obtain a DEFUZZIFIED VALUE (Inferred VALUE) from this set, there is no unique way if the membership function does not exhibit a unique maximum as is the case in the figure. **At this juncture, the sanctity of the purely fuzzy (non-arithmetic) operations is sacrificed by taking recourse to some arithmetic akin to the SOP formula of the INS as follows.**

(i)   Taking the mean of the flat maximal region of the membership function, Mean of Maximum (MOM) which is tantamount to the use of SOP formula with uniform weight for each of the values in the maximal flat region.

(ii) Taking the centroid of the entire membership function which is also tantamount to the use of the SOP formula- Center of Gravity (COG) or Center of Area (COA) method.

In this chapter we have seen the way in which SINGLE entities, constants, vectors and functions are represented in various ways. Funtions representations and their transforms into other domains have been also seen. In all these, the way in which the INS structure is reflected has been illustrated. In most cases, the radix-like term and its power term (position indicator) are evidently as in the INS. However, in the special case of generalized function, the shifted function appears in place of the power term, which means that the positioning indicated by the shifted argument is natural unlike in the other cases where positioning is by a power term of the radix-like term.

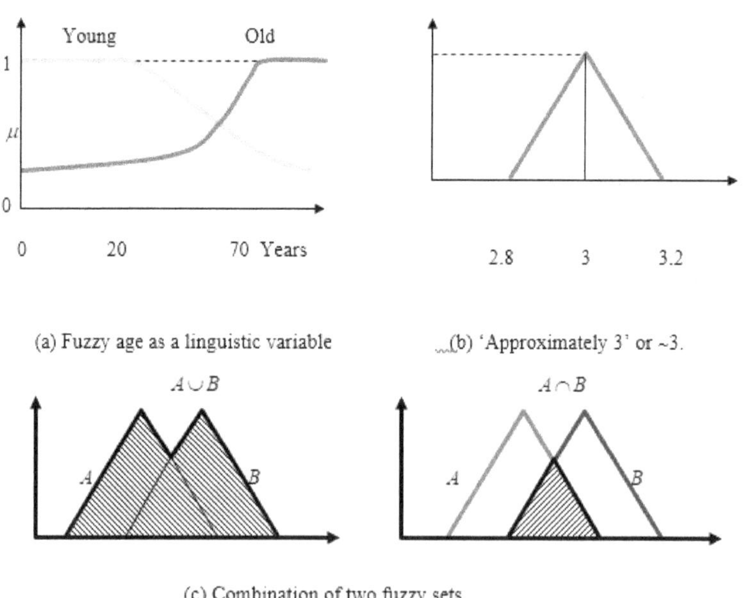

(a) Fuzzy age as a linguistic variable   (b) 'Approximately 3' or ~3.

(c) Combination of two fuzzy sets

Figure 5.9 Examples of fuzzy sets

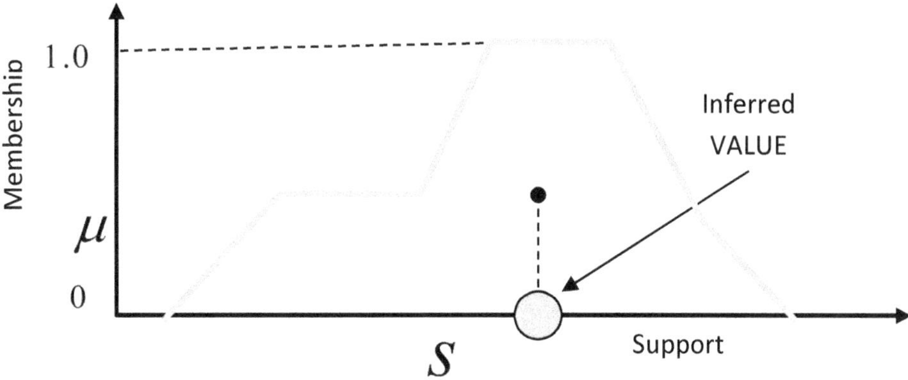

Figure 5.10 A fuzzy quantity represented by a fuzzy set S with membership function μ

Relationships Between Pairs of Arrays of Discrete Entities and of Functions

6.1 Two Vectors

In the previous chapter we have seen the way in which SINGLE entities, constants, vectors and functions are represented in various ways and the shades of similarity with the INS. Two arrays appear in representation of a single entity, one to denote the entity and the other as one which gives a meaning, like scaling, to the former.

Now, we will consider pairs of SEPARATE vectors and functions which appear in SOP fom for evaluation of their VALUES which are measures of different MUTUAL features between the pairs. The individual arrays may arise in various separate ways and one may be often interested in determining a quantitative measure of relationship between the two. Such situations often arise in scientific investigations or engineering applications as will be clear from the presentation that follows.

Scalar Product and Orthogonality

Given two n-vectors $\mathbf{u} = \begin{bmatrix} u_1 & u_2 & \cdots & \cdots & u_n \end{bmatrix}^{\mathrm{T}}$ and $\mathbf{v} = \begin{bmatrix} v_1 & v_2 & \cdots & \cdots & v_n \end{bmatrix}^{\mathrm{T}}$, the SOP formula in gives their scalar product

$$R = \mathbf{u}^{\mathrm{T}} \mathbf{v} = \begin{bmatrix} u_1 & u_2 & \cdots & \cdots & u_n \end{bmatrix}^{\mathrm{T}} \begin{bmatrix} v_1 \\ v_2 \\ \vdots \\ \vdots \\ v_n \end{bmatrix}.$$

The two vectors above are said to be *orthogonal* if the scalar product $R = 0$. All the related study of the vectors follows. The use of the SOP for evaluation of the relationship between the vectors can thus be traced back to the INS as explained in Chapter 5.

6.2   Two Functions

6.2.1   Inner Product

At first let us understand the term PRODUCT which we use in different contexts. Literally the term 'product' means manufactured goods, result of a process, produce such as a crop, invention, artifact, result outcome, consequence, effect, and so on.

In the case of arithmetic of two single numbers it is the result of multiplication. This is applicable in the context of evaluation of entities such as area   ($length \times breadth$),   physical   work   done ($force \times distance\ in\ the\ direction\ of\ the\ force$), Torque, Couple, Moment, etc. ($voltage \times current$)

In the case of assessing the total importance of a set of entities the product of weight or importance of the individual entities is summed up into a single value.

In the case of a trade transaction the total value of the transaction is taken as the SOP of rates and the corresponding amounts of goods in terms of numbers, units of volume or weight. The total value shopping bill is a simple example of SOP application in our daily lives.

In the case of resultant forces in the determination of entities such as center of gravity or centroid (first moment), moment of inertia (second moment) the average of the sum pf products is a well known formula.

In the event of calculating power and energy we take the products- of Voltage and Current, and of Voltage and electrical charge respectively.

6.2.2 Abstract Entities as Functions

There are situations in which we require some quantitative measure of an abstract entity such as *effectiveness of cooperation* between two parties in collaboration working to achieve a common goal. In such situations, if the collaborative process involves activities on each part of the parties over the period of the common project symbolically denoted by functions of time $f(t)$ and $g(t)$, the inner product of these two functions over the period provides a measure of effectiveness of collaboration or cooperation between them. The inner product is the result of integration of the product of the two functions over the period and is a single number. In the case of two physical processes such as voltage and current in an electrical system it is the total net energy or work done. **In general, the value of the inner product is a measure of synergy between two functions.**

Functional analysis is an extension of vector analysis in infinite dimensions if functions are regarded as arrays of sampled values of the function; the number samples approaches infinity as the sampling interval approaches zero. As in the case of the vector space, in the function space, the energy of a function is calculated by its norm over its interval of definition, $\|f(t)\|$ which on a normalized base can be taken directly as mean and this is called the Root Mean Square (RMS) value, especially in the context of electrical waveforms. The inner product of two functions $f(t)$ and $g(t)$ denoted as

$$\langle f(t), g(t)\rangle = \langle g(t), f(t)\rangle$$

is calculated as in the case of vectors but in continuous domain and the final summation is replaced by integration as shown in Figure 6.1. Two functions are said to be orthogonal if the inner product is zero. In the case of electrical systems the functions may represent voltage and current respectively and if their inner product is zero, they produce no work and are also said to be in quadrature. This sense of angle in electrical systems is understood with all vectors termed as *Phasors.*

The inner product is useful in evaluating the relationship between two processes $f(t)$ and $g(t)$ working together. This may be a measure of synergy or cooperation. It may be the content of one function extracted from another in its own form. If sinusoidal functions are used for one, the inner product gives the harmonic components of the other. If orthogonal polynomials such as Jacobi, Chebychev, Legendre, Laguerre polynomials are used, we get the corresponding expansions of the other function. If Poisson functions are used in place of one, we get the Poisson moment functionals (PMF) of the other function. In the case of characterizing a function in other forms such as Poisson moment functionals (Saha and Rao, 1983), we take the products of the given function with members of the Poisson function set. With large λ the Poisson function becomes a narrow pulse which may be viewed as a wavelet. Modulating function concept of Shinbrot (1957) predates the idea of wavelets.

The Poisson Moment Functionals (PMF) of a signal $f(t)$ are real numbers $f_k^{t_0}$, $k = 0, 1, 2, \ldots$, generated by

$$f_k^{t_0} = \int_0^{t_0} f(t) p_k(t_0 - t)dt \quad ; k = 0, 1, 2, \ldots$$

$$p_k^t = p_k(t) = t^k e^{-\lambda t} / k! ; \ \lambda(\text{real}) > 0.$$

These can be physically generated by applying the signal $f(t)$ to a Poisson filter chain with identical elements, each of which has a transfer function $1/(s + \lambda)$. The output of the

k th stage happens to be the Poisson moment functional. By letting $\lambda = 0$, we generate successive higher order integrals which are also useful in signal characterization for system identification (Saha and Rao, 1983). Use of these in the identification of continuous-time models of dynamic systems eliminates the need to directly differentiate process signals and gives rise to algebraic equations in which the parameters of the continuous-time model appear to be estimated. With $\lambda = 0$, initial conditions appear in the process of evaluation and they too can be estimated to become the estimated state variables of the system, thus enabling us to simultaneously estimate the state and parameters of the system. Usually, a large value of λ, larger than the largest of time constants, renders the effect of initial conditions vanish. But, the bandwidth of the Poisson filter becomes very wide making it pass all noise components into the estimation process.

Extraction of particular forms of components from the content of a function

In another case of practical importance, we wish to determine the content of one function *in its form* in another function and vice versa. In this case also the inner product can be used. An example is the case determining the harmonic components in a periodic function in which case the period of consideration will be the period of the function. In other words, it is a process of extracting the harmonic components from the given periodic function.

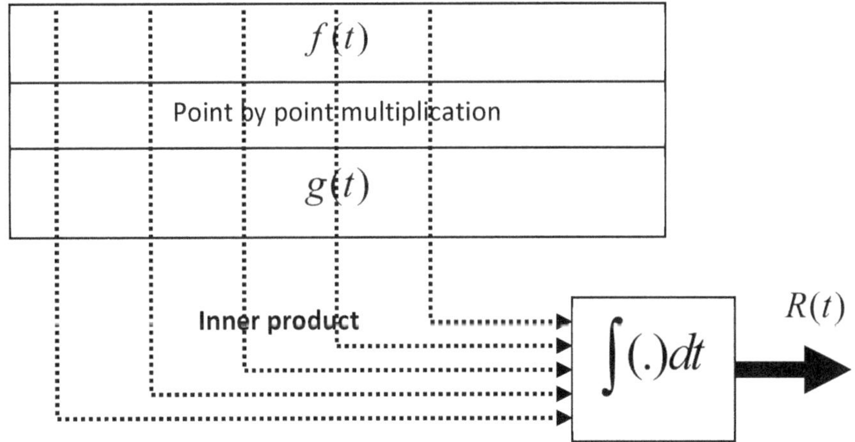

Figure 6.1 Inner product of two functions

Transformation of the domain of a function

In the definition of integral transforms of functions of time, we use a Kernel function of two variables- one being time and the other a variable in whose domain we wish to obtain the transform. For example, in the case of the Laplace transform of a function of time, we use a Kernel of the form $K(s,t)$ resulting in the Laplace transform in the domain of the complex variable s . The process of integration **winds up the time side of the process leaving the result as a function of** s **alone.**

The framework of the inner product can be extended to the process of convolution involving two functions. Convolution is nothing but inner product evaluation in which one of the functions is delayed with respect to the other. The result for various values of delay over the whole interval becomes a new function of time. The process applied to advanced argument in one of the functions takes us towards the concept of correlation functions.

6.2.3   Inner Product: Contexts of Application

(a)   Deterministic environment

Singletons: Area, Work done, Moment, Torque, Couple, etc.

Weightage/Importance

Total cost: Sum of (Rates x Quantities)

Resultant force, Center of gravity, Moment of inertia

Power and energy

Cooperation

Relatedness/Dependence

Extraction

Synergy

Positioning kernel in representations and integral transforms

Moment

Moment functional

Convolution integral: Shifted inner product

(b) Uncertain Environment

Expected value

Correlation: Auto and cross: Shifted inner product

Now let us consider two functions $f(t)$ and $g(t)$ over an interval which can be the normal interval $[0,1]$ without loss of generality. That is, the scaling of results, such as averages, over the interval will not take place due to the unit length of the interval and our discussion will continue without the distraction due to the scale factor.

The two functions may be relevant to the study of some real-world problems such as the following.

1. Rainfall as one function and the water demand as the other both over one year for seasonal management of drainage and irrigation for a seasonal crop.

2. Voltage as one function and current as the other in an electrical energy system

3. Work schedules of two parties meant to collaborate with each other.

Generalizing the case of two vectors discussed above to continuous time of infinite dimensional case, the following analogue of the scalar product shown above becomes

$$R = \int_0^1 f(t)g(t)dt \ .$$

Symbolically the above is written as

$$R = \langle f(t), g(t) \rangle \ .$$

6.3   Interpretation of the Value of the Inner Product

The value of the inner product as is another *avataar* (incarnation) of the INS SOP formula may be seen in the three practical situations:

1. The effectiveness of water supply for the crop by rain

2. The Work done over the interval by electricity where the current and voltage are applied

3. The effectiveness of mutual collaboration/cooperation between the two parties

The value p of the inner product is indicative of the *relationship* between the two functions. We can understand what it means if $p = 0$; the crop should not depend on the rainfall because it gets no water from it. Likewise, in the electrical system no work is done and the electric current flowing through the network merely heats the circuit without doing any useful work. In the last case the cooperation between the collaborating parties is nil due to total failure of communications; when any party calls the other for discussion there will be no response. This situation where the inner product of two functions over an interval is zero is known as orthogonality.

A relationship moved towards orthogonality and the synergy between the parties is reduced due to the reduction in the common working days from five to two. In the year 2021 during the period of restrictions due to the pandemic Covid-19, this matter has been corrected and the working week days in some of the countries in the Middle East are synchronized with the rest of the world.

The sense of orthogonality may be like each entity casting its shadow on the other in light falling vertically from above each individually. The term *independence* is different from orthogonality despite the use of the inner product formula as it arises in the study of suspected dependence between processes. There should be some inner causality relationship between processes to term this as dependence.

Mathematically features such as synergy, dependence, and coupling have the inner product as the evaluation formula but the use of the name for the relationship depends on an understanding of the relationship between the processes.

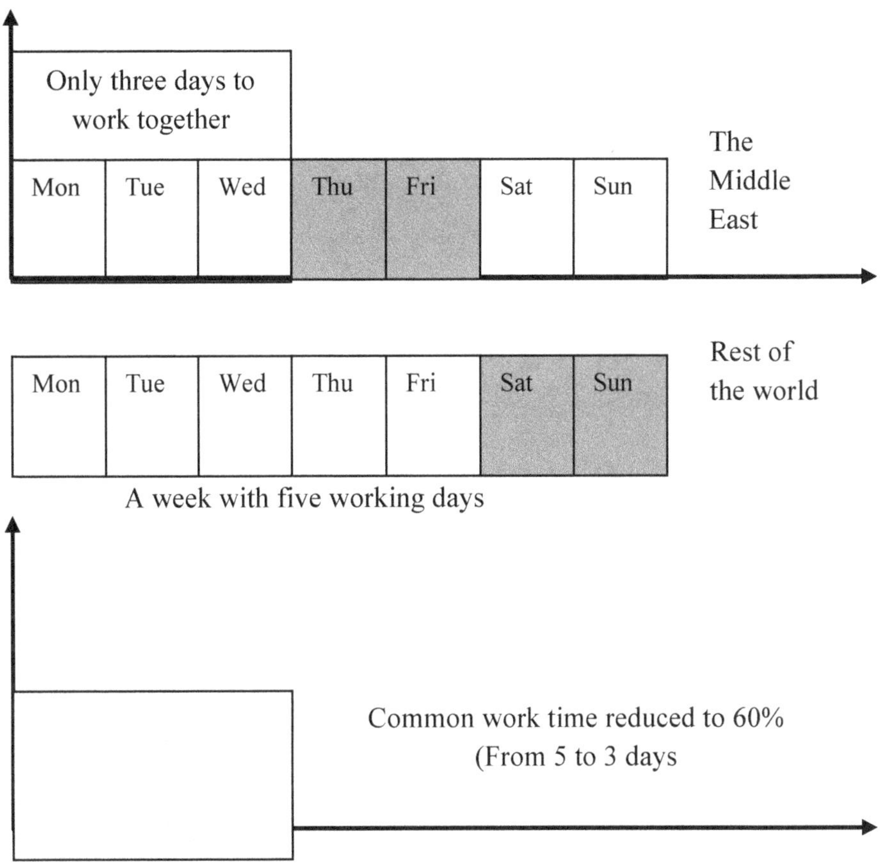

Figure 6.2 Effectiveness of collaboration between two parties with different weekly work-days

In the above three examples, rainfall and water required for crop and work schedules of the collaborating parties are not mutually dependent. In the electrical system however, the current will depend on the parameters of the circuit. In a purely reactive circuit- inductive or capacitive- the current is in *quadrature* with the applied voltage which means orthogonality. Many situations in real life represented by functions with data observed on different processes, the pairs are studied for correlation between them. The inner product or similar forms for calculating the relationship between processes are like SOP formula of the INS for evaluation.

The feature of orthogonality is exploited for some advantages in mathematical applications. For example, functions are expanded in series of orthogonal functions as basis. The advantage of choosing an orthogonal basis for approximating functions lies in the fact that the error in any term does not affect other terms at all. We can go on adding higher order terms to the series approximation without the need to look into what happens to the previously included terms. The discrete Fourier series is one example.

If the orthogonal basis is A periodic function $f(t)$ over the period $[0,1]$ is expanded in the above series as

$$\left[f_0 + f_1 \cos \Omega t + \overline{f_1} \sin \Omega t + f_2 \cos 2\Omega t + \overline{f_2} \sin \Omega t + \cdots \right],$$

The coefficients of the series are evaluated as the following inner products:

$$f_0 = \langle f(t), 1(t) \rangle,$$
$$f_1 = \langle f(t), \cos \Omega(t) \rangle,$$
$$\overline{f_1} = \langle f(t), \sin \Omega(t) \rangle,$$
$$f_2 = \langle f(t), \cos 2\Omega(t) \rangle,$$
$$\overline{f_2} = \langle f(t), \sin 2\Omega(t) \rangle,$$
$$\ldots \text{and so on}$$

In general the k-th order frequency term

$$f_k = \langle f(t), \cos k\Omega(t) \rangle,$$
$$\overline{f_k} = \langle f(t), \sin k\Omega(t) \rangle$$

The inner product in evaluating the coefficient of terms in the orthogonal series approximations is a measure of the net useful energy over the period of definition of the functions. The multiplication of two functions gives rise to a product or square, in the sense of energy. The process is akin to obtaining the amount of the component under consideration at its maximum level while *minimizing the energy of the residue or error*. Therefore, **it is an optimal method of obtaining the coefficients of an orthogonal series expansion-least squares optimization**

In general in the inner product

$$R = \langle f(t), g(t) \rangle,$$

R may be viewed as the component of $f(t)$ in $g(t)$ and vice versa as in any orthogonal series. Of course there is only term $f(t)$ and or $g(t)$ and the inner product gives the component of one in the other. When such evaluation takes place for the component of $f(t)$ in $g(t)$, the residual error

$$e_g = g(t) - p\, f(t)$$

will be orthogonal to $g(t)$ and is the smallest error in the sense of least squares. Likewise, if evaluation takes place for the component of $g(t)$ in $f(t)$, the residual error

$$e_f = f(t) - p\, g(t).$$

The value of p remains the same in both cases. Thus the inner product acts as a scooping formula for the component of a function from inside another function. The scoop is optimal in the sense of least squares. The scooping is commutable.

Please note that we consider functions $g(t)$ and $f(t)$ over a normal interval, $T = 1$ and also having unit norm over the interval to be free from the distracting scale factors. Otherwise, averaging and normalizing needs to be done after taking the inner product.

6.4   Least Squares Curve Fitting

Given a set of experimental data $\{t_k, f_k\}, k = 1, 2, \ldots$, one may wish to fit an nth degree polynomial function of the form

$$g(t) = \sum_{k=0}^{n} a_k t^k$$

optimally in the sense of least squares; that is, by minimizing the sum of squares of the error over the interval of data.

The error at t_k

$$e_k = g(t) - f_k$$
$$e_k^2 = \left[g(t_k, a_0, a_1 \cdots a_n) - f_k \right]^2$$
$$E^2 = \sum e_k^2 = \left[g(t_k, a_0, a_1 \cdots a_n) - f_k \right]^2$$

The conditions for optimality are

$$\frac{\partial E^2}{\partial a_k} = 0,$$

a set of equations with unknowns a_k . The problem is then to solve the set for the unknown coefficients a_k to obtain the desired optimal polynomial. In order to readily see the procedure, let us consider the problem in a simplest form as follows:

$$g(t) = a_1 t,$$

a straight line passing through the origin; that is, $a_0 = 0$ .It is straight line starting the origin.

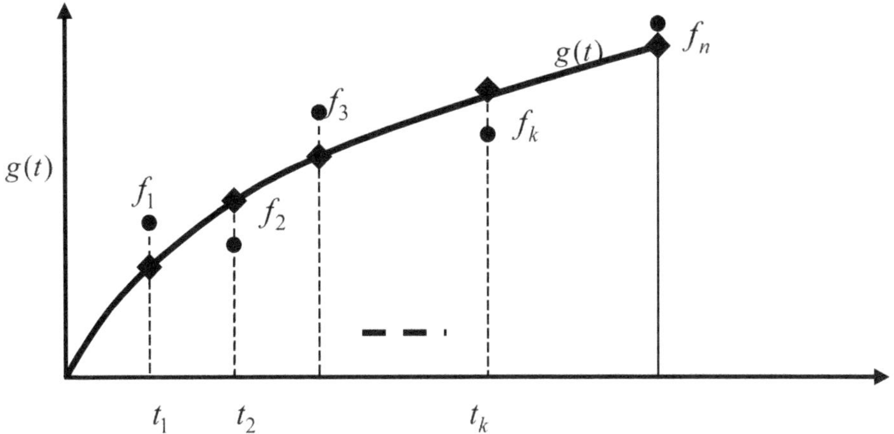

Figure 6.3 Fitting a function to a set of observations using least squares

We can look at this problem in the following way: The fitting function is in the form $g_0(t) = t$ and we have to determine the amplitude of $g_0(t)$, which is a_1, to fit optimally.

In this case,:

$$E^2 = \sum e_k^2 = \sum \left[a_1 g_{0,k} - f_k \right]^2 = \sum \left[a_1^2 g_{0,k}^2 - 2a_1 g_{0,k} f_k + f_k^2 \right].$$

The condition for minimum of this is

$$\frac{\partial E^2}{\partial a_1} = 2a_1 \sum g_{0,k}^2 - \sum 2g_{0,k} f_k = 0,$$

giving

$$a_1 = \frac{\sum g_{0,k} \cdot f_k}{\sum g_{0,k}^2}$$

Notice that the numerator is the inner product of the form of the fitting function with the function $f(t)$ denoting the data. The denominator is the norm of the basis function $g_0(t)$. If $g_0(t)$ were of unity norm, the denominator would have been unity. In Fourier coefficient evaluation the denominator is unity since each of the terms in the basis is normal. The core aspect of this method is that it involves inner product which is akin to the INS.

A statement as follows can therefore be made in the case of a pair of functions: ***"The least squares scoop of a function from another function on the same interval is the inner product of the two functions over the interval. The magnitude of the least squares scoop of the other function from the first function will be the same on the interval."*** Considering a pair of functions $f(t)$ and $g(t)$, the component of $f(t)$ in $g(t)$ is given by $\langle f(t).g(t) \rangle$. Due to commutability the same will be the component $g(t)$ in $f(t)$. The property of orthogonality reduces the error in the cases of orthogonal series expansions, content extraction, and fitting given data to a function in a chosen form- usually a polynomial. In all these cases, the innate property of the inner product is to leave error which is minimal in the sense of least squares. A residual error that is orthogonal to the fitted function also implies that it is optimal in the sense of least squares.

Array-Pair Evaluation with Shifted Argument

7.1 Convolution and Filtering

When the SOP formula of the INS is extended to integrals akin to the inner product with any of the functions in the pair is subjected to a shift τ that varies from 0 to t, the constant R evolves into a function

$$R(t) = \int_0^t f(\tau)g(t-\tau)d\tau$$

or

$$R(t) = \int_0^t f(t-\tau)g(\tau)d\tau.$$

Integrals as the above occur in many practical applications such as convolution, correlation, filtering, etc. We will now look into these integrals that may be viewed to have evolved from the SOP formula of the INS.

A linear dynamic system with impulse response function $g(t)$ driven by an input signal $f(t)$ gives the output as

$$R(t) = \int_0^t f(\tau)g(t-\tau)d\tau$$

or

$$R(t) = \int_0^t f(t-\tau)g(\tau)d\tau.$$

If the Laplace transforms of $f(t)$, $g(t)$ and $R(t)$ are $F(s)$, $G(s)$ and $R(s)$ respectively,

$$R(s) = F(s)G(s) = G(s)F(s)$$

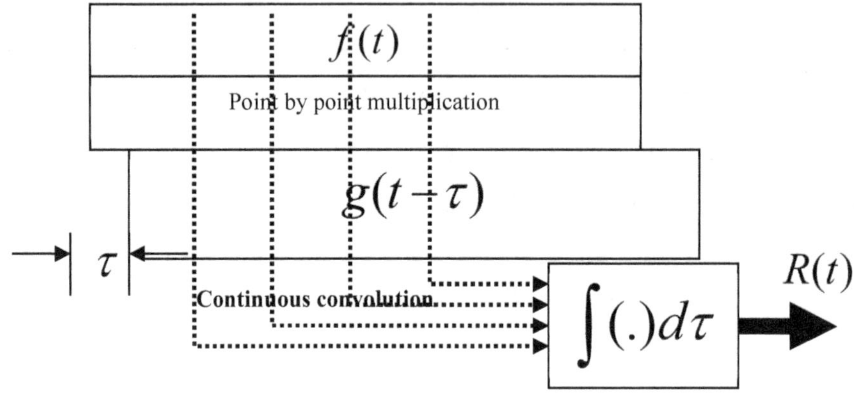

Figure 7.1 Convolution in Continuous Domain

which is simply algebraic multiplication. In time domain the relation is to be evaluated by its dual- convolution.

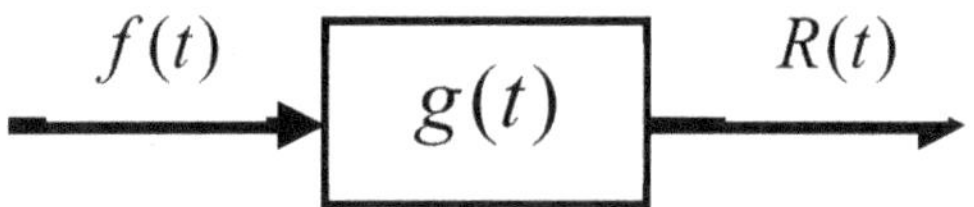

Figure 7.2 A linear time-invariant dynamic system

The same output is obtained when a system with impulse response $f(t)$ is driven by an input signal $g(t)$. The output of a linear system is given by the above integrals which are called convolution integrals.

This is the case of application of the SOP evaluation formula in the integral form due to continuity of the data; one of the functions is shifted by τ and the evaluation takes place by varying the shift from 0 to t . The result will now be a function of t .

A filter is a dynamic system with desired properties to filter the input signal. The dynamic system to function as filter of unwanted signal components and to pass the desired frequency components is set up in signal processing applications. Based on the requirement to have the desired components the dynamic system is chosen to have its frequency response characteristic to perform the function. If a low pass property is needed, the dynamic system which is usually an electrical circuit is designed with low pass property and attenuating property at high frequencies. In the normal operation of a given system if high frequency noise is expected, a low pass filter is used. Likewise if a band pass property is desired, a dynamic system whose frequency response characteristic has cut off frequencies corresponding to the pass band is designed. Figure 7.3 shows the frequency response characteristic of a band-pass filter

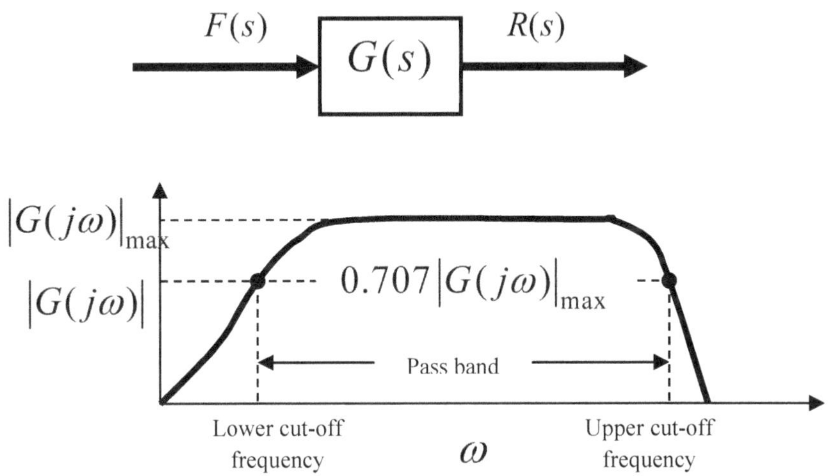

Figure 7.3 A band-pass filter

Therefore, the above dynamic system may be regarded as a filter in appropriate circumstances and the integrals can be regarded as filtering equations. Recall the inherent least squares feature of optimality in the inner products; these properties do apply to the filtering case. Therefore the above filtering is in the sense of least squares.

7.2   Correlation Functions

7.2.1 Auto-correlation Function

Given a real function $f(t)$ the autocorrelation function of $f(t)$ is

$$R_{ff}(\tau) = \lim_{T \to \infty} \frac{1}{2T} \int_{-T}^{T} f(t) f(t-\tau)\, dt ,$$

where τ is a time delay parameter. It shows how random a signal is by showing the influence on average, of the value of $f(t-\tau)$ on the value of $f(t)$ by exhibiting a significant peak at τ. For a perfectly random signal the autocorrelation function $R(\tau)$ becomes an impulse at τ. In experiments where a random signal is required, say, in dynamic system identification, an ideally random signal with a peak only at $\tau=0$ cannot be generated since its the Fourier transform

$$S_f(\omega) = \frac{1}{T} \int_{0}^{T} R_{ff}(t) e^{-j\omega t}\, dt$$

shows uniformly the same constant value of the signal at all frequencies from 0 to ∞, implying unlimited power in the signal. Figuratively such a signal is called a *white* signal. The term *white noise* is used to refer to an input test signal for revealing all modes of a dynamic system in the identification process. While studying noise, signals that are disturbances to the main process signals, the attribute *noise* is understandable. But in identification of dynamic systems when such a signal is used we must refer to it as *illuminating* signal since it throws light on all the critical frequencies of the test system. Only in white light can one see all colors. In practical experiments since it is impossible to generate an ideally white or random test signal, it is general practice to generate pseudorandom signals in the form of square waves. These are called pseudorandom binary signals (PRBS). The autocorrelation function of a PRBS shows peaks at periodic intervals. If the period of this repeatedly occurring signal is designed to be long enough to contain the impulse response function of the dynamic system under test decays within the period of the PRBS, such a signal even if not ideally random, will be adequate to capture all the critical frequencies of the system under test. Therefore, calculation of autocorrelation function is of practical importance and the definition of this function clearly suggests that it is akin to the convolution process inherent in the INS. It is a process of evaluation of the integral as SOP with varying values of τ followed by averaging over the entire time record.

7.2.2 Cross Correlation Function between Two Functions

Cross correlation is a measure of similarity of two functions as a function of displacement of one relative to the other. It is the inner product or dot product evaluated at a variable shift between the functions.

$$R_{fg}(\tau) = \int_{-\infty}^{\infty} f(t)g(t-\tau)dt$$

It is not commutative.

$$R_{fg}(\tau) \neq R_{gf}(-\tau)$$

7.2.3   The Array-Pair Structure of the INS

The array pair structure as the common thread between the INS and many mathematical methods of representation and evaluation of relations. It plays an important role in connecting the calculus of dynamic systems with algebraic operations. It helps in quantifying some of the abstract features such as cooperation between collaborating parties in their joint work. It helps in establishing some quantitative relationships which are not clear for direct causal analysis. Situations in which correlation analysis helps are some social processes. If such processes are monitored in time, correlation measure between them would pave way to making assessments of significance and for possible empirical relationships. In scientific studies experimental investigations may first indicate some dependence between two variables, which may possibly culminate in establishing useful quantitative relations for application. In the early years of scientific studies, phenomena were merely observed and by consistent continued efforts useful relationships among the various factors causing the phenomena were evaluated to ultimately become the so-called empirical formulas.

A mathematically meaningful pair of arrays, one for representation of an entity and the other a scaling array that is associated with the representation with a sequence of weights, was first observed in the INS and it was introduced to the West by the book of Fibonacci in 1202 CE. The scaling array is not shown while writing a number in the Indian decimal system. Its subsequent generalization as a number system on any base greater than or equal to 2 became possible. The number system with the smallest base, the binary system became the breath and pulse of our modern systems of computing and communication. Addition of two numbers in this representation is direct with carryover to higher locations since the number with base r has digits d_k, each limited to $r-1$ in its location and when the result of addition or any other arithmetic operation gives rise to a result greater in value than $r-1$, the excess has to be carried over to the next location $k+1$. Multiplication of two numbers in the INS representation is by convolution of the discrete numbers in the representation arrays.

It was long after the INS came into use, the mathematics of quantities in more than one dimension, that is the so-called vector analysis, appeared. The coordinate system proposed in 1637 CE by Rene Descartes which is now known as the Cartesian system became useful to illustrate the location of points in two dimensions. The extension of this to higher dimensions gave rise to the so-called Euclidean space. In this space the coordinate axes are orthogonal to each other. The analysis of vectors in two dimensions with the Argand diagram evolved into higher dimensional methods with the introduction of vector matrix methods in the 1850s by James Sylvester and Arthur Cayley. The relationship between two vectors $\mathbf{u}$ and $\mathbf{v}$ is evaluated through the scalar product $\mathbf{u}^T\mathbf{v}$ or $\mathbf{v}^T\mathbf{u}$. Following the method of evaluation in the INS framework, the energy of a vector $\mathbf{u}$ is calculated by $\|\mathbf{u}\|$ the square root of the product of the vector by itself by, i.e., $\mathbf{u} = \mathbf{v}$ as its own weighting array. The vector elements are all directly denoted by their respective values in the format of the INS; there is no pair array for each element and the arithmetic follows the rules of the INS. If $\mathbf{u}^T\mathbf{v}$ or $\mathbf{v}^T\mathbf{u}$ becomes zero, the vectors are said to be orthogonal to each other.

In the case of a random quantity, one array represents the various possible values of the quantity and the other array denotes the probability values corresponding to the values of the random quantity. The SOP in this case gives the expected value of the uncertain quantity.

A fuzzy quantity is represented as a fuzzy set. Each value of support is associated with a value of membership function. In order to obtain a crisp value out of the result of a combination of fuzzy sets for further use the evaluation is based on the concept of centroid of the area under the membership which is given by the SOP process. The impact of the INS on the world of modern mathematics is summarized in Tables 7.1 (a) and (b).

The weighted sum feature in neural networks also should remind us of the INS.

Table 7.1(a) The array pair structure of the INS in mathematical formulas for representation

S. No.	Case	Array/function (1)	Array/function (2)	Evaluated entity	Remarks
0	**INS**	$\{d_k\} \to \{0, 1, 2, \dots 9\}$	10^k	**Value:** $\displaystyle\sum_{k=-\infty}^{k=\infty} d_k 10^k$	**INS**
			Progeny of the INS		
1	General number system on any positive integer base $r \geq 2$	String of digits $\{d_k\}$	Positional Values r^k	Value of the Number: $\displaystyle\sum_{k=-\infty}^{k=\infty} d_k r^k$	General number system. Digits are all positive integers $0, 1, 2, \dots, (r-1)$, r being the radix or base
2	n-dimensional vector **X** in Euclidean space	Elements $x_k, k = 1, 2, \dots, n$ in orthogonal coordinates	**x**	Norm of the vector $\|\mathbf{x}\| = \sqrt{\displaystyle\sum_{k=1}^{n} x_k^2}$	
3	Function $f(t)$ over $[0,1]$	$f(t)$	$f(t)$	$\|f\| = \sqrt{\int f^2(t)dt}$	Effective or RMS value
4	Fourier expansion of a periodic function over a normal interval	$f(t)$	$e^{jk\Omega t}$	$a_k = \int f(t)e^{jk\Omega t}\, dt$	Evaluation of Fourier coefficients
5	Laplace traznsform	$F(s) = \mathcal{L}f(t)$	e^{-st}	$F(s) = \displaystyle\int_0^{\infty} f(t)e^{-st}\, dt$	Transformation from t domain to s domain

Contd...

S. No.	Case	Array/function (1)	Array/function (2)	Evaluated entity	Remarks
6	General integral transform	$F(s) = \mathcal{L} f(t)$	The kernel $K(s,t)$	$F(s) = \int f(t)K(s,t)dt$	$K(s,t)$ literally means 'at t'
7	General inverse transform	$f(t) = \mathcal{L}^{-1} F(s)$	The inverse kernel $K^{-1}(s,t)$	$f(t) = \int F(s)K^{-1}(s,t)ds$	$K^{-1}(s,t)$ literally means 'at s'
8	z-transform of $f(t)$ sampled at interval of T	$f(kT), k = 0,1,\ldots$	z^{-k}	$F(z) = \sum f(kT)z^{-k}$	z^{-k} is the position operator of shift of k intervals
9	Random quantities	$x_i,\ i = 1,2,\ldots,n$	$p_i,\ i = 1,2,\ldots,n$	Expected value $$\mu = E(X) = \sum_{1}^{n} x_i p_i$$	
10	Fuzzy set S	Support array $s_i,$ $i = 1,2,\ldots,n$	Membership function values $\mu_i, i = 1,2,\ldots,n$	Inferred/ defuzzified value s	Methods: Mean of maximum (MOM) of membership, COG

Table 7.1(b)　The array pair structure of the INS in mathematical formulas for evaluation of relationships

S. No.	Case	Array/function (1)	Array/function (2)	Evaluated entity	Remarks
11	Vectors of the same dimension	Vector $\mathbf{x}$	Vector $\mathbf{y}$	Inner product/ Scalar product $$p = \mathbf{x}^{\mathrm{T}}\mathbf{y} \text{ or } p = \mathbf{y}^{\mathrm{T}}\mathbf{x}$$	Orthogonality implies that $$p=0$$
12	Functions over an interval	$f(t)$	$g(t)$	Inner product $$p = \langle f(t), g(t) \rangle = \langle g(t), f(t) \rangle$$	Orthogonality implies that $p=0$. A relationship can have different names: Synergy, Effectiveness of joint action, etc.
13	Convolution	$f(t)$	$g(t)$	$$R(t) = \int_{0}^{t} f(\tau)g(t-\tau)d\tau \text{ Or}$$ $$R(t) = \int_{0}^{t} f(t-\tau)g(\tau)d\tau$$	Convolution appears in many situations: Multiplication of two numbers, response evaluation for a linear dynamic system, filtering, etc.
14	Auto-correlation function of a signal $f(t)$	$f(t)$	$f(t)$	$$R_{f}(\tau) = \lim_{\tau \to \infty} \frac{1}{2T} \int_{-\tau}^{\tau} f(t)f(t-\tau)dt$$	
15	Power spectrum of a function	$R_{ff}(t)$	$e^{-j\omega t}$	$$S_{f}(\omega) = \frac{1}{T} \int_{0}^{\tau} R_{ff}(t)e^{-j\omega t} dt$$	This is Fourier transform formula
16	Cross correlation	$f(t)$	$g(t)$	$$R_{fg}(\tau) = \int_{-\infty}^{\infty} f(t)g(t-\tau)dt$$	$R_{fg}(\tau) \neq R_{gf}(-\tau)$ Not commutative

Representation of Polynomials in MATLAB

MATLAB is a very well known computational package with toolboxes for special subjects. The format of representation of polynomials is identical with that of the INS. A n-th degree polynomial in x is considered in decreasing powers of x and is represented as an array of the $n+1$ coefficients of its terms. If a term does not exist, its coefficient is set as 0 keeping the array length as $n+1$. That is, Zero coefficients must be marked as zero entries. Thus,

$$p(n) = a_n x^n + a_{n-1} x^{n-1} \cdots + a_1 x + a_0$$

is representated as an array

$$\begin{bmatrix} a_n & a_{n-1} & \cdots & a_2 & a_1 & a_0 \end{bmatrix}$$

of $n+1$ elements corresponding to the coefficients associated with decreasing powers from left to right. For example, [3 2 1] represents $P\circledR = 3x^2 + 2x + 1$ whereas [1 0 0] represents $P(x) = x^2$. The array of coefficients is in the same mathematical sense as the array of digits in the INS. The polynomial is understood as a SOP of the two arrays

$$\begin{bmatrix} a_n & a_{n-1} & \cdots & a_2 & a_1 & a_0 \end{bmatrix} \quad \text{and} \quad \begin{bmatrix} x^n & x^{n-1} & \cdots & x^2 & x & 1 \end{bmatrix}$$

including the order of arrangement of the powers from left to right.

Concluding Remarks

Let us consider some questions instead of making assertions. The reader may find these questions useful in coming to any conclusion.

- Isn't the INS the only system among the many candidates in ancient times that has a mathematical form with two arrays and a value formula in SOP form?

- Was the link between the INS and the many mathematical concepts which have a common structure as listed in Tables 7.1(a) and 7.1(b) noticed in any of the mathematical developments prior to the arrival of them in the West?

- What could be in the mind of Pierre Simon Laplace who made such admiring remarks on the INS and wondered how some of the great minds in the West missed the idea? Laplace had not seen the surge of computers during his time as we see today. Even in the case of Einstein, we cannot elaborate on his admiration in clear terms.

- How do we explain the conspicuous presence of the same structure in all the mathematical concepts centuries after the arrival of the INS?

- Is it right then to say that it is because the INS became so commonplace its structure did not warrant any reference?

- Even in the representation of polynomials in the well-known mathematical computations package and its associated tool boxes, the orders of the terms increase from left to right and the polynomial happens to be result of SOP operation on the terms of the two arrays, just as in the case of INS. The array of powers is not shown in the representation and it is implicit in both cases.

- Since no reference to the INS structure is evident in the works of the mathematicians who first proposed these concepts, did the ideas occur in total independence of the INS concept?

- Is the use of the structure of the INS, especially the binary number system in computers which became ubiquitous in our lives- regular and professional, solely responsible and adequate to validate the claim that the INS has changed the world?

- We use only the decimal system and the Hindu-Arabic numerals from our early childhood but we are seldom concerned about the origin of the numbers and methods of calculation. We take them for granted and we never cite the originators. After all when we calculate the total shopping bill in the same way.

References

1. Amir Aczel, *Finding Zero- A mathematician's odyssey to uncover the origins of numbers*, St. Martin's Publishing Group, 2015.

2. Charles Seife, *Zero-Biography of a dangerous idea*, Viking Adult, 2000

3. Dines Chandra Saha and Ganti Prasada Rao: Identification of Continuous Dynamical Systems - The Poisson Moment Functional (PMF) Approach, LNCIS Vol. 56, Springer Verlag, Berlin, 1983.

4. Ganti Prasada Rao and D.A. Rutherford,: Approximate reconstruction of mapping functions from linguistic descriptions in problems of fuzzy logic applied to systems control, *Cybernetics and Systems*, Vol.12(3), pp 225-236, Sept. 1981.

5. Ganti Prasada Rao and Darwish M.K.F.Al Gobaisi: A pattern of expressions from Indo-Arabic arithmetic to the mathematics of signals and systems, *International Conference on Mathematical Theory of Control*, Bombay, India, December 1990.

6. Ganti Prasada Rao and H. Unbehauen, Identification of continuous-time systems, IEE Proceedings Control Theory and Applications, Vol.153, N0.2, March 2006.

7. Ganti Prasada Rao, Piecewise constant orthogonal basis functions and their application to systems and control, Springer Verlag, Berlin, Heidelberg, 1983.

8. Ian Gullberg, *Mathematics from the Birth of numbers*, W.W. Norton & Comapany, 2007

9. Georges Ifrah, *The universal history of numbers*, John Wiley, 2000

10. J. W. McCrindle, *Ancient India as described by Megasthenes and Arrian*: Maluka Publisjing LLC, 2008

11. Robert Kaplan, The Nothing that is : A natural history of Zero, Oxford University Press, 2000

12. Shinbrot, M.: 'On the analysis of linear and nonlinear systems', *Trans. ASME*, 1957, **79**, pp. 547-552

13. Terry Jones, The story of One; Documentary (https://vimeo.com/56113926)

Index